THE

MARINE AQUARIUM

Comprehensive coverage, from setting up an aquarium to choosing the best fishes.

Dick Mills

CONTENTS – *Practical Section*

First edition for the United States and Canada published in 2005 by Barron's Educational Series, Inc.

First published in 2004 by Interpet Publishing

All inquiries should be addressed to:
Barron's Educational Series, Inc.
250 Wireless Boulevard
Hauppauge, NY 11788
www.barronseduc.com

International Standard Book Number 0-7641-2987-2

Library of Congress Catalog Card Number 2004101703

Printed and bound in China
9 8 7 6 5 4 3 2 1

Author

Dick Mills, author of many aquarium books, is a former editor of *Aquarist and Pondkeeper* magazine and also a vice president of the Federation of British Aquatic Societies. He has kept aquarium and pond fish continuously for the last 40 years, but also finds time to travel, when – as if by accident – he always manages to come across a public aquarium or two.

Other contributors:
Colin Grist
Andrew Caine
Steve Halls

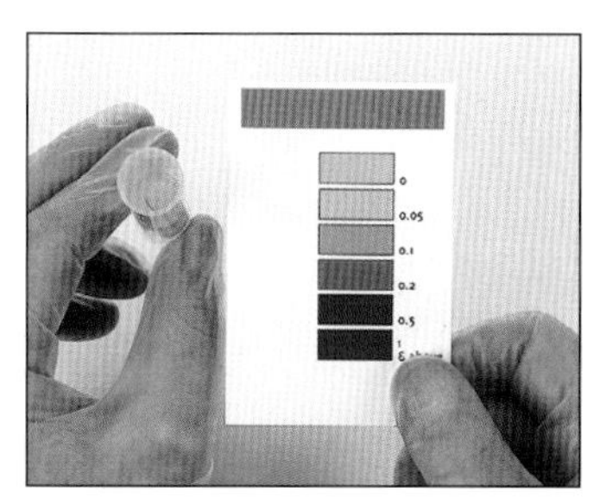

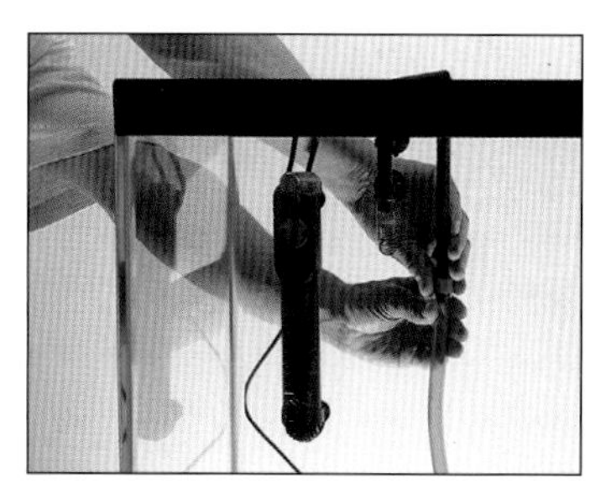

CONTENTS – *Fish Profiles*

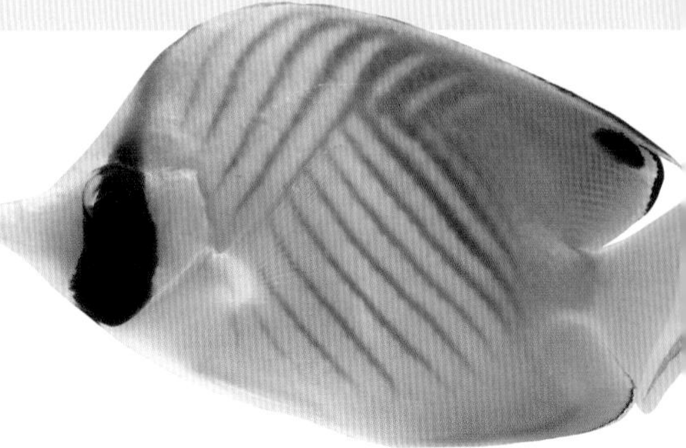

Part Two: Fish profiles *86-201*

PART ONE

The marine aquarium is different

If your fishkeeping experience has been confined to tropical freshwater fish, the biggest difference you will find when you start keeping marine species (apart from the salt) is the fact that the number of fishes you can keep in a tank of any given size is far fewer than you have been used to.

A second difference, especially in a basic marine setup, is the absence of aquarium plants. Although synthetic corals may be a reasonable substitute as a means of providing sanctuary for the fishes, they are nothing more than static objects. However, once you have gained experience, you can progress from a simple marine setup to a reef aquarium. This will enable you to enjoy the vibrant background provided by macro-algae, living corals, and other invertebrate life.

The major impact of setting up a marine aquarium is that it places exacting responsibilities on its owner. Successful marine fishkeeping is in direct proportion to the amount of effort you are prepared to put into maintaining optimum conditions.

Although marine fishkeeping may be considered a relatively recent introduction within the fishkeeping hobby, there is more pioneering work to be done, particularly in the area of captive breeding. Experience built upon the efforts of others in the past, together with the technological advances in aquarium equipment, have provided us with a much firmer foundation from which to advance. Every marine fishkeeper has something worthwhile to contribute.

Marine fishkeeping is no longer in the "hit and miss" or "try and hope" categories of years gone by. Modern air transportation systems ensure that livestock arrives in far better condition, aquarium equipment is completely reliable, and there is a wider range of suitable aquarium foods available. And, with the vast amount of information available in publications and on the Internet, both the experienced fishkeeper and the novice have a wealth of backup advice and guidance at their disposal to help them in the quest for success.

Bringing the natural world into your home

Switch on the television or go on holiday to a sun-drenched island and the chances are that you will see (or be offered) an underwater journey on which you will encounter the brilliant colors and dazzling shoals of fish as they cruise around the coral reefs.

Modern day technology, while allowing you to view and visit such sights, also enables you to create a living souvenir in your own home. The fast jet transport that so effortlessly carried you away is also likely to be bringing thousands of fish to your aquatic dealer. No longer quite so dramatically stressed by the protracted journey, these living jewels are soon thriving in almost identical conditions to those of their native homes as they await your interest. But is it as easy as it sounds?

All it takes is to provide the fish with the same conditions as they would enjoy in nature – excellent water conditions, enough living space and a steady supply of nutritious foods. Fortunately, all the old problems encountered in the early days of marine fishkeeping have been resolved. Crucially, the ability to create synthetic seawater has removed the risk of introducing pathogens into the aquarium by using natural seawater. Modern equipment and practical knowledge make maintaining the water in optimum condition an easy task, while a greater appreciation of the lifestyle and needs of the animals themselves means that they will not be doomed to a short aquarium life as was the case in earlier times. A further encouragement (should you need one) is that since most marine aquarium fishes are caught in shallow waters from coastal fringing reefs, the fishes that you enjoyed seeing in the wild will be readily available for your home aquarium, too.

Left: *Looking down on the coral reefs may be a beautiful experience in itself, but to the marine fishkeeper nothing beats getting down among the fishes or, by following the steps described in this guide, bringing them into your own home.*

The bountiful oceans

The Indo-Pacific Ocean is the natural home of a very large proportion of marine fishes suitable for the aquarium. Despite the huge areas of very open water, most are collected in shallow waters from local reefs.

Fish collected from almost anywhere on this map will all be satisfied by one set of water conditions – so stable is their world-ranging environment.

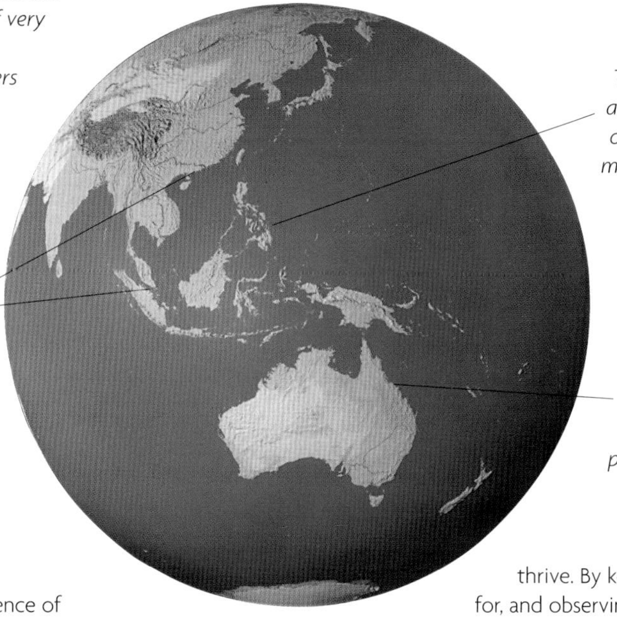

The Philippines are a rich source of fishes for the marine aquarium.

Singapore and Hong Kong are important hubs of the fish collecting and exporting trade.

The Great Barrier Reef is visited by thousands of tourists each year – all potential marine fishkeepers.

If you add to these reassuring facts the existence of suitable aquarium materials well able to withstand the effect of proximity to corrosive seawater, then the possibility of a successful marine aquarium becomes a reality rather than a dream.

Although it is tempting to assume that recreating a coral reef in the home is within everyone's reach, remember that there is still a lot to learn before you can provide just the right environment in which the underwater panorama of your imagination will thrive. By keeping, caring for, and observing the fishes in your aquarium, you could be making a contribution to their conservation. The more we know about the fascinating animals in our charge, the more we can perhaps replicate their needs for captive culture. In this way, we could reduce the need to catch them from the wild, denuding their natural homes and, at the very least, sparing them the long flights that, no doubt, they find as exhausting and tedious as we do.

Seawater – simulating the real thing

As 71% of the Earth's surface is covered by seawater, you might expect it to be one of the most convenient commodities to obtain. Unfortunately, it is not often suitable for the domestic marine aquarium, for several reasons.

For most aquarists it is totally impractical to make regular trips to the coast to collect and transport large quantities of natural seawater. Secondly, relatively few marine aquarists have the luxury of living in the tropics and being able to collect local water. For most, collecting cooler local water and converting it to "tropical water" can present a number of problems. Warming up natural seawater could cause plankton either to die off, creating potentially toxic conditions, or rapidly multiply and use up vital oxygen. In addition, coral reef fishes are unlikely to have the natural means to fight off disease pathogens that occur in water from other parts of the world. Should you wish to establish an aquarium to house fishes from your local shores, then it is acceptable, but not always convenient, to use seawater from the same locality.

Another important point to consider is the difficulty of finding a source of unpolluted natural water. The increasing volume of seagoing commercial traffic and effluent from industrial activity around the world means that coastal waters are not likely to be entirely pure, to say the least!

It is clear from this that using natural seawater in a home aquarium can generally create more problems than it is worth.

Synthetic seawater

The best way of providing suitable water for the marine aquarium is to use an artificial mix that is carefully balanced and combined with good quality tapwater so that the final composition approximates as closely as possible to natural seawater. These commercially available mixes of dry salts, minerals, and essential trace elements are sterile and should

***Right:** The constant movement of the sea caused by tides and wind-driven waves means that it is always saturated with oxygen.*

The "salt mix" of seawater

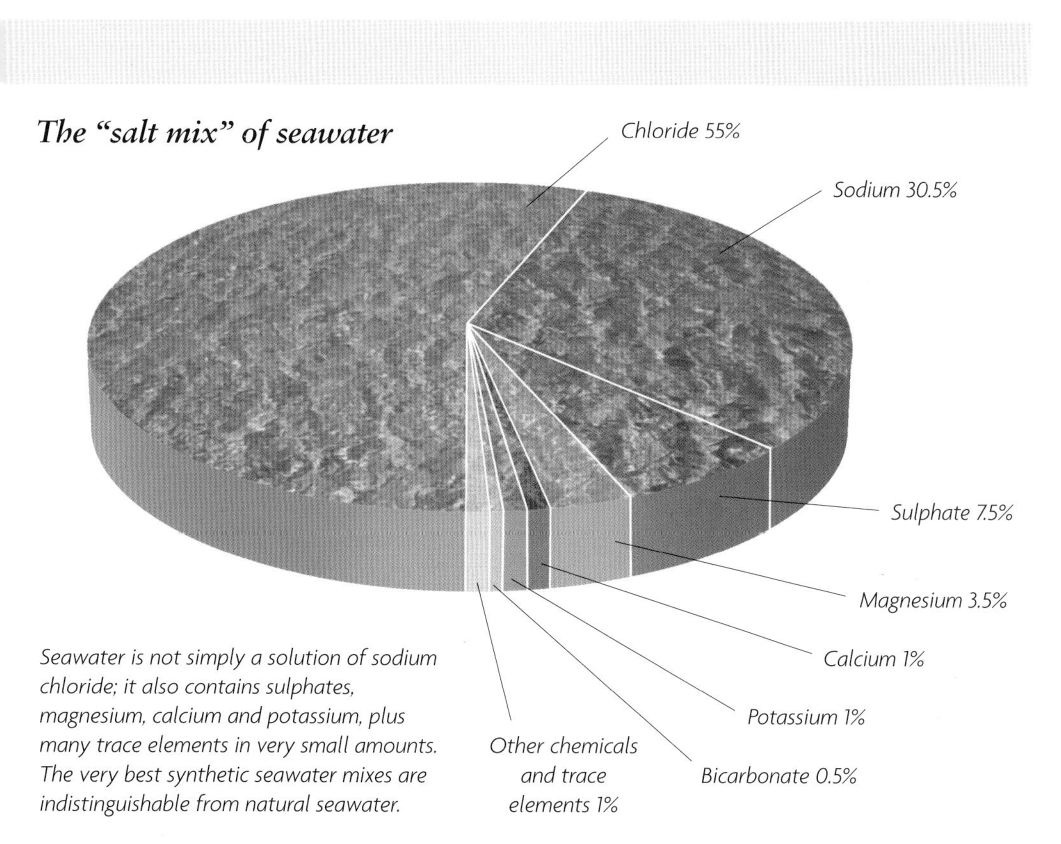

Seawater is not simply a solution of sodium chloride; it also contains sulphates, magnesium, calcium and potassium, plus many trace elements in very small amounts. The very best synthetic seawater mixes are indistinguishable from natural seawater.

be manufactured from scientific-grade materials, so that there is no risk of inadvertently introducing diseases and other problems into the aquarium. Also, because you only purchase the dry mix and add it to freshwater when needed, it is very convenient in terms of carrying and storing.

Modern mixes are so exact in their formula and manufacturing process that you no longer need to use up an entire package at once in order to produce a given volume of water with the desired specific gravity (or density) and correct chemical composition. These days you can use up part of a package and still be sure of obtaining the correct chemical composition regardless of what specific gravity you wish to attain. (Reseal partly used bags and store in a cool, dry place.) This offers a great deal of flexibility and is an ideal solution for preparing small amounts of seawater for partial water changes. This is the medium your fishes and invertebrates have to live in, so quality is essential.

Living in a salt solution

The key difference between marine fish and freshwater fish, both coldwater and tropical, is the level of salt found in the water they inhabit and their ability to survive in their environment. While some freshwater fish may adapt to life in saltwater, and some marine fish will survive at low salinity levels, their inability to cope with the level of salts both in their body and in the surrounding water is one major reason why marine fish and freshwater fish will not generally survive in each other's environment. The process by which the salts are absorbed and excreted from the fish's body and the way in which they control their salt levels is known as osmoregulation.

How osmoregulation works

In all fish there is a natural difference between the level of salts found in the fish's body fluids and the level of salts in the surrounding environment, whether it is freshwater or saltwater. Only very thin membranes, such as the gills, separate the differing levels, so there is a tendency for water and salts to flow constantly into and out of the fish. This two-way movement is caused by diffusion and osmosis, natural processes that occur wherever two solutions of different concentrations are separated by a partially permeable membrane. The salt ions move from the more concentrated solution through the membrane and into the weaker solution; this is diffusion. The water molecules will move in the opposite direction and dilute the stronger solution; this is osmosis.

Marine fish differ from their freshwater relatives in the way they balance the levels of salts and fluids in their bodies. The bodies of freshwater fish contain a higher level of salts than the surrounding water.

Above: *Marine fishes are well adapted to thrive in their highly salty environment. They can absorb water and excrete the salt to maintain their internal bodily fluids at the right concentration for life.*

There is a tendency for salts to be lost through the tissues as water flows in. To combat this, freshwater fish have extremely efficient kidneys that are able to excrete water very quickly, while retaining the salts in the water by reabsorbing them from the urine. Additionally, freshwater fish have specially structured gills that allow them to retain salts circulating in the blood that have been ingested with food.

In marine fish the situation is reversed. Their environment has a higher salt level than is found in their body fluids. As a result, they face a constant battle against dehydration, as water is lost to the surrounding seawater and salts are retained in the body. Marine fish deal with this situation by consuming large amounts of water and excreting very little urine. Only a few salts are absorbed from the seawater they drink, and special chloride cells in the gills actively excrete salt.

Freshwater osmoregulation

Saltwater osmoregulation

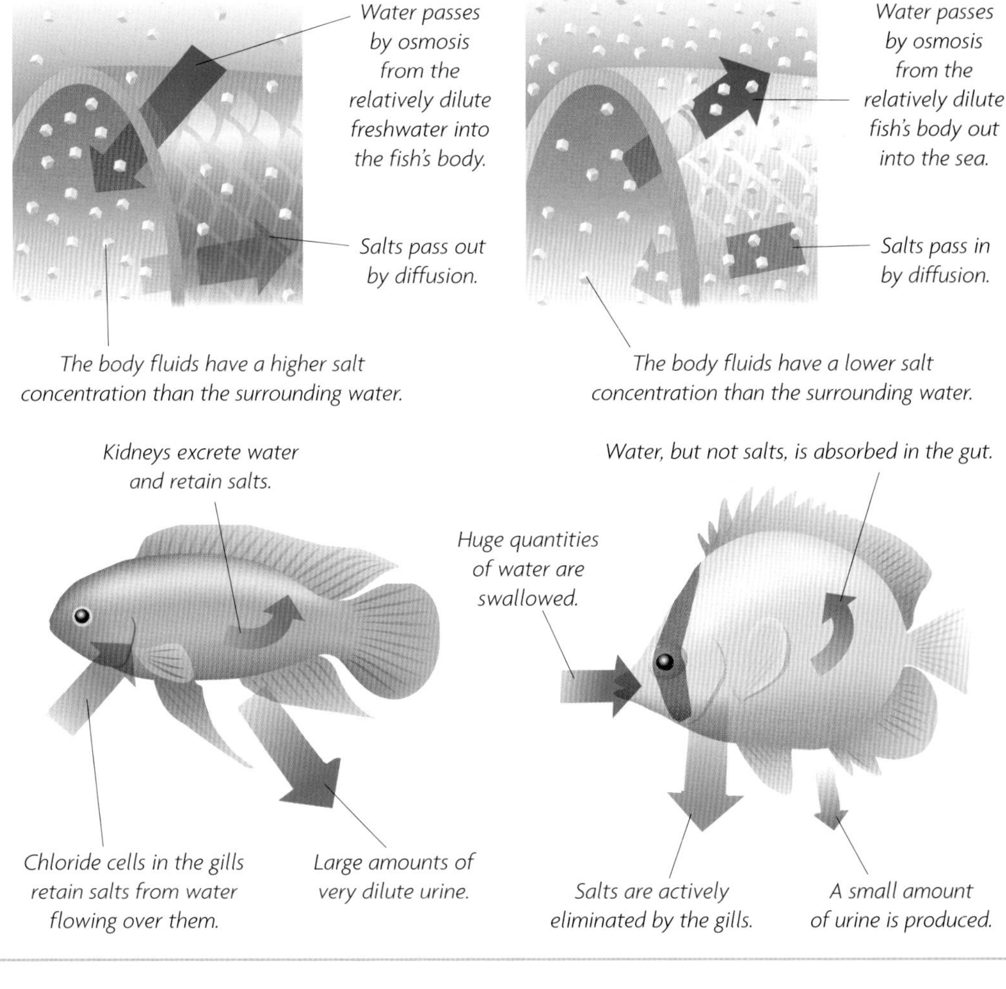

Salinity, specific gravity, and pH

The salinity of water is a measure of the total amount of dissolved salts it contains; it is usually quoted as gm/liter (parts per thousand). Within the fishkeeping hobby, however, it is more usual to talk in terms of the specific gravity (S.G.) of water, since this is more easily tested and clearer to understand.

Specific gravity is simply a ratio of the weight of a water sample compared with the weight of an equal volume of distilled water at 39°F (4°C), which is assigned a specific gravity of 1. Since adding salts to water increases its weight as well as its salinity, the two scales are directly comparable. Thus, a salinity of 35 gm/liter is equivalent to a specific gravity of 1.026 at 59°F (15°C), measured by means of the traditional floating hydrometer or the easy-to-use swing-needle types.

Seawater generally has an S.G. of between 1.023 and 1.027, depending on the location. Once properly acclimatized, many fish and invertebrates seem happy to accept a constant S.G. in a fairly wide range. For most purposes, a reading between 1.021 and 1.024 is acceptable. The key factor is to keep the S.G. as constant as possible. By using a hydrometer, take regular readings and ensure that they do not vary by more than one point (between 1.022 and 1.023, for example). Although this seems a minor shift, it is still many times that which occurs in even the most extreme circumstances in any one natural location.

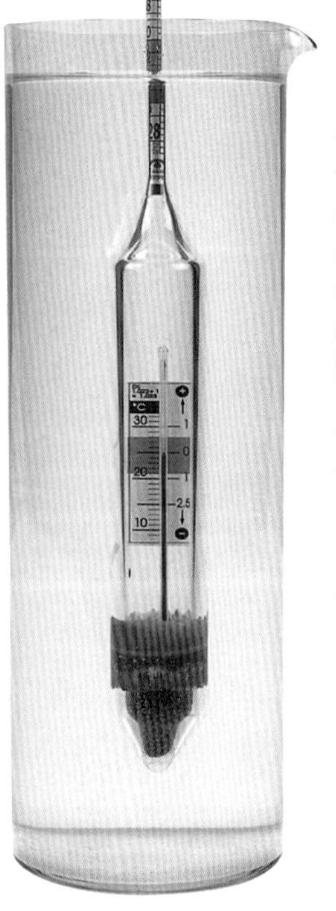

Left: *The correct specific gravity (S.G.) is indicated when the chosen value coincides with the meniscus (water level) as shown on the colored part of the stem.*

In an established tank, the main cause of changes in S.G. are evaporation and the addition of new stock. Over a period of time, the water level in the tank tends to drop as freshwater evaporates, and thus the salt water become progressively more saline. Compensate for this by regularly adding a small quantity of *fresh* water. It is not a good policy to wait until a large amount (more than 0.5 liter/ approximately a pint, say) is needed, as the change in salinity will harm delicate invertebrates and algae. Two common and dangerous sequences of events can lead to a continual increase in the S.G. of the water in your tank. In the first scenario, you allow the water level in the tank to drop through evaporation and then decide to carry out a partial water change. Having drained out 10-20% of the water, you replace exactly the same amount with newly made up seawater, plus extra seawater to make up the evaporation loss. Since the reduced volume in the tank was more concentrated, this will cause an increase in S.G. Alternatively, the water level drops and you then add more stock (and its attendant seawater), thus filling the tank and also raising the S.G. Be sure to guard against these pitfalls.

How pH is measured

The pH scale registers the acidity or alkalinity of water. The scale is logarithmic, so that each unit change in pH, say from 7 to 8, is a ten times change, while 7 to 9 is a hundred times change, etc. This is why a sudden change in pH is very stressful and harmful to fish.

Water molecule

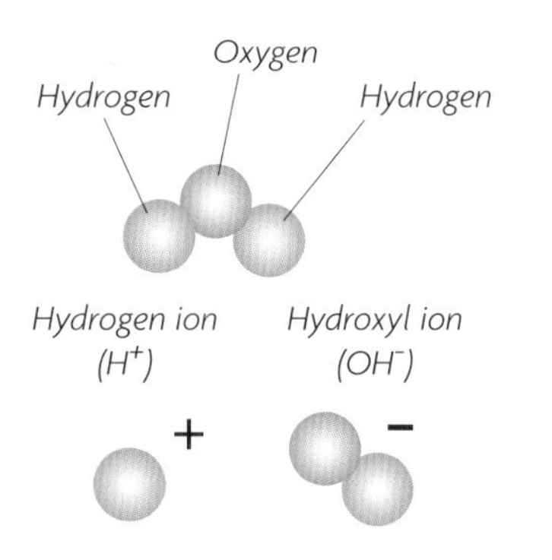

Water (H_2O) is made up of positively charged hydrogen ions (H^+) and negatively charged hydroxyl ions (OH^-). The pH level is a measure of the ratio of these two ions in a body of water. Acidic water has more hydrogen ions; alkaline water more hydroxyl ions. Neutral water has an equal number of both.

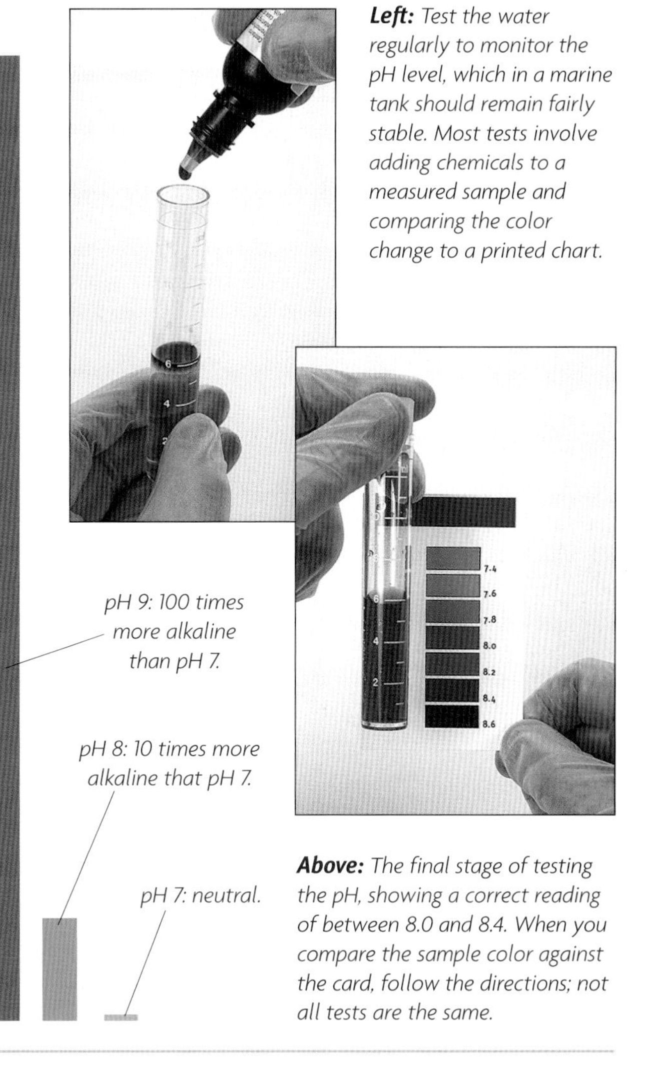

Left: *Test the water regularly to monitor the pH level, which in a marine tank should remain fairly stable. Most tests involve adding chemicals to a measured sample and comparing the color change to a printed chart.*

Above: *The final stage of testing the pH, showing a correct reading of between 8.0 and 8.4. When you compare the sample color against the card, follow the directions; not all tests are the same.*

The nitrogen cycle

The nitrogen cycle is an extremely important biological process on the reef – and in all ecosystems, for that matter. It is an important purification system, but it is more than that, because it is also the source of primary food production, without which nothing on earth would survive.

How the cycle works

The nitrogen cycle is the "driving force" that removes harmful nitrogenous compounds that gradually build up in the aquarium water, starting with free ammonia (NH_3) and ionized ammonia (NH_4^+). Fishes and invertebrates produce ammonia as a waste product from their metabolic processes. This is added to the ammonia produced by bacteria "working on" other waste materials in the aquarium, such as uneaten food and feces. Ammonia is highly toxic to fishes and invertebrates; if it is not removed, or converted into other less harmful substances, your aquarium subjects will soon perish. Fortunately, in Nature's Grand Plan, a substance that is poison to one living organism is food for another and so there is a natural way of dealing with ammonia removal.

Aerobic (oxygen-loving) bacteria, such as *Nitrosomonas* species, convert ammonia to nitrite (NO_2), a slightly less toxic substance, but one that is still dangerous to fishes and invertebrates. A second group of bacteria, such as *Nitrobacter* species, transform the nitrite to nitrate (NO_3), which is a much safer substance but one that can still cause some problems if allowed to build up too much. It is not fully understood to what level nitrate can cause problems for aquarium subjects, although it seems to be detrimental to delicate fish species and some invertebrates. The important thing is that nitrate does not occur on the natural coral reef other than in the smallest quantities, so we should not make compromises but strive to keep it to an absolute minimum in the aquarium. Nitrate is a primary nutrient for algae and the mainstay of primary food production right at the beginning of the food chain from which all other living organisms benefit. (It is worth noting that undesirable algae flourishes in the presence of excess nitrates, as well as phosphates.)

Left: *Marine ("macro") algae, such as the* Caulerpa *sp. shown here, will thrive in a marine aquarium. The nitrates produced as part of the nitrogen cycle are absorbed as food by these and other less welcome forms of algae.*

How the nitrogen cycle works

Nitrogen is recycled in the natural world through the digestion of nitrogen-containing proteins by animals and the action of bacteria in the environment. Here is how the cycle works in the marine world.

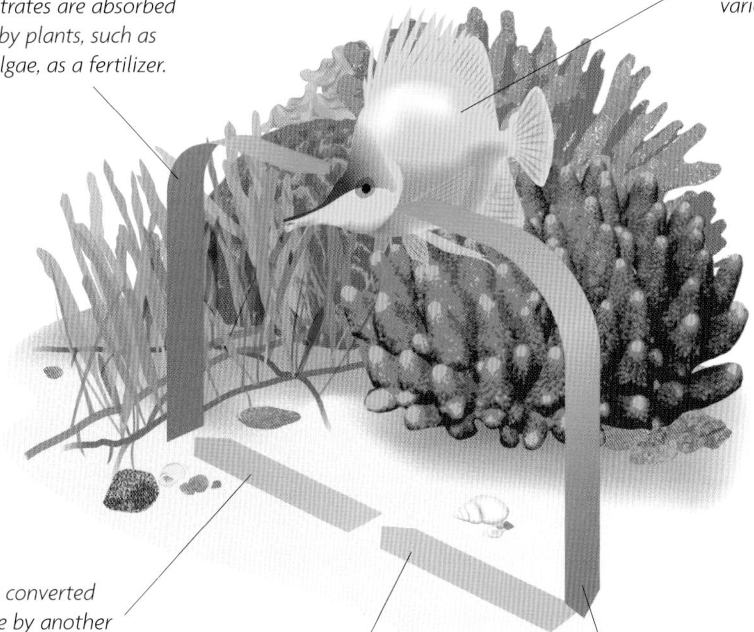

Fish and other marine creatures digest and metabolize protein as part of their varied diets.

Nitrates are absorbed by plants, such as algae, as a fertilizer.

Nitrite is converted into nitrate by another group of bacteria called Nitrobacter.

Ammonia is converted by Nitrosomonas *bacteria into nitrite. These bacteria thrive in oxygenated conditions in the substrate and on the surface of rocks.*

The main waste product of protein digestion is ammonia. Fish excrete this in urine and directly from the gills. Ammonia also builds up as feces, plant matter, and uneaten food decay.

Protein skimmers

Most fishkeepers consider protein skimmers to be essential for the long-term success of a marine aquarium, but what do they do and how do they work? Originally designed for the sewage treatment industry, protein skimmers, or foam fractionizers as they are sometimes called, remove harmful organic substances from the aquarium before they can cause the water quality to deteriorate.

A protein skimmer works on the principle that dissolved matter, including bacteria and plankton, as well as organic material resulting from fish waste and food, will adhere by surface tension to the outer layer of air bubbles. The rising column of air bubbles reaches the top of the skimmer's "reaction chamber," where it falls away, leaving the protein waste to collect at the top of the skimmer. This protein manifests itself as a thick, yellowish, smelly liquid.

A triple-pass protein skimmer

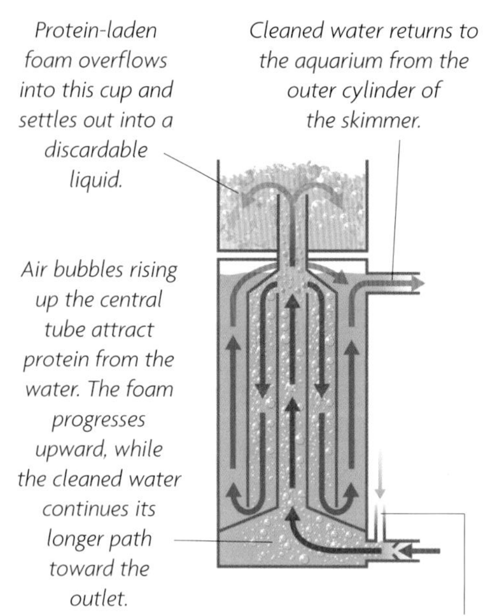

Contact time

Certain factors play an important role in the efficiency of a protein skimmer. One of them is contact time, the period during which air and water are mixed together in the skimmer chamber. The longer this period is, the more effective the skimmer becomes. The triple-pass, Berlin-type system represents the best available.

Types of protein skimmers

Protein skimmers are available in three main forms. **Air-operated protein skimmers** are mounted inside the aquarium or sump and are generally better suited to smaller aquariums. **Venturi skimmers** are high-powered workhorses, operated by a water pump that takes water either from the aquarium or the sump and pushes it into the bottom of the skimmer through a venturi tube that draws in air to create a mass of bubbles. **Turbo skimmers** have air injected directly into a turbo pump with a modified 12-blade impeller that chops up the air and water into a super-fine concentrated mix. Turbo skimmers, especially those using Berlin technology, have set new standards for removing protein from aquariums. **Convergent-divergent flow skimmers** have a unique method of mixing air and water, while still using the turbo method of air injection.

An advanced turbo skimmer

This advanced skimmer maximizes the contact time between the tank water and a constant stream of fine air bubbles. Protein wastes stick to the bubble surfaces and can be collected and removed as the bubbles collapse.

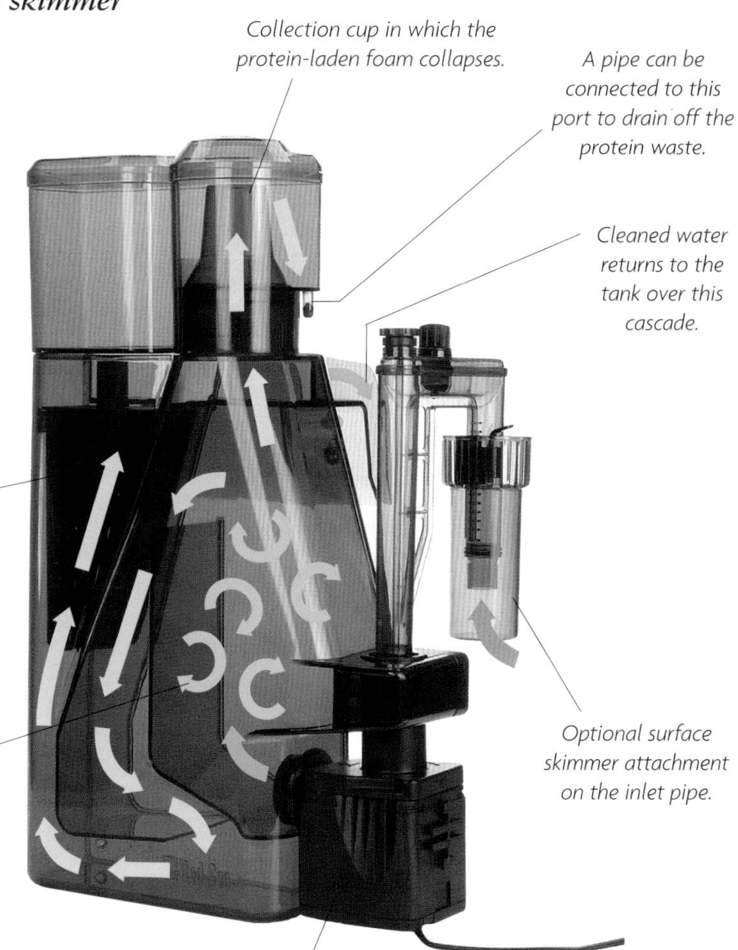

External filters

Although protein skimming is a type of filtration, it is important enough to be discussed in its own right. But, generally speaking, "filtration" refers to mechanical, biological, or chemical filtrations, which are all used in various ways to provide marine fishes with stable, healthy water.

The filter is the life-support system of any aquarium. It not only prevents debris from remaining within the aquarium by physically removing it into a filter chamber or some other container (mechanical filtration) but also breaks down the toxic ammonia compounds released into the aquarium by fish excreta and the decay of excess food (biological filtration). If carbon or phosphate removers are used, the filter also performs chemical filtration. Over the next few pages, we look at filtration systems and consider their advantages and disadvantages.

External power filters

Also known as canister filters, external power filters are fairly large filters situated outside, usually below, the aquarium. Canister filters enable you to use different filter media (usually two or three) in one filter. By varying the filter media, you can incorporate all three forms of filtration – mechanical, biological, and chemical.

Ideally, the filter has an initial prefilter foam to remove particulate waste. The foam is usually followed by a filter media with a high surface area, such as sintered glass, which is quickly populated by millions of beneficial bacteria that break down ammonia and nitrite into less harmful nitrate (see page 16). Finally, carbon may be used to remove any slight discoloration from the water and give the aquarium an extra "sparkle."

An external canister filter

These plastic tubes carry water to and from the aquarium.

Shut-off taps disconnect the filter without water spillage.

An electric water pump is housed in the top part of the filter.

Incoming water passes upward through the filter media, packed in a plastic basket inside the canister. Maintain water flow at all times to prevent the media turning anaerobic, as would quickly happen following any failure of the power supply.

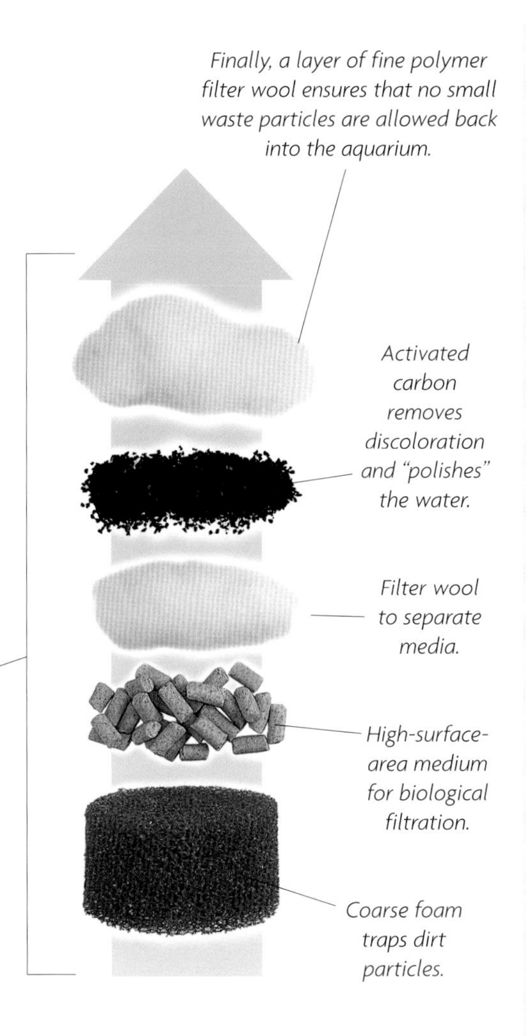

Algae filters

As excess nitrate/phosphate levels encourage algal growth, it seems a sensible plan to turn this fact around and deliberately use algae to remove these unwanted "pollutants." A separate shallow container can be placed within the water flow pumped from the aquarium. In it, and under fairly bright lighting, a healthy crop of algae can be grown without it interfering with the designed beauty of the main tank. By regularly harvesting the algae, nitrates and phosphates can be effectively removed from the aquarium water; as this main algal growth develops it outcompetes any would-be algal growth within the main aquarium in the race for nutrients in the water. One drawback of this system is that any herbivorous fish in the main aquarium served by an algae filter will be denied their necessary green diet; arrange for these fishes to get their "green" food by other methods. To be effective, such filters need to be of considerable size, which may be problematical when it comes to disguising them, lest they detract from the inherent beauty of the otherwise well-presented main aquarium.

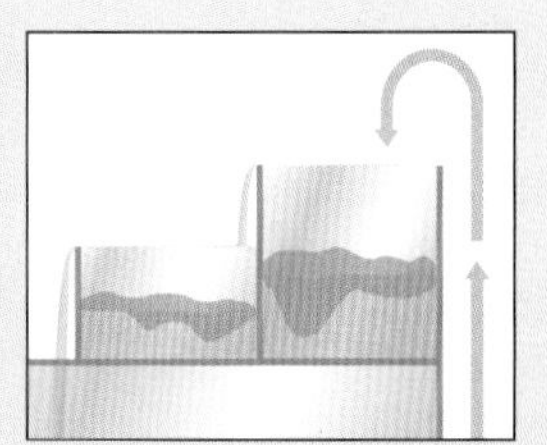

Right: *This shows the basic principle of an algae filter, but the shape and scale must be matched to the display aquarium.*

Fluidized bed filter system

Fluidized bed units take biological filtration out of the tank and supersede outdated substrate systems. Silica sand is held in suspension in aquarium water pumped through an external cylindrical container. Because the sand granules are free to move, and are not packed together, they offer a much larger surface area for bacterial colonization, and this results in a much more efficient conversion of ammonia through nitrite and into nitrate.

In some models, an incorporated carbon filter cartridge removes any coloration from the water before it returns to the aquarium. The fluidized bed system is oxygen-consuming and water emerging from such units should be given every chance to re-oxygenate before returning to the main aquarium. This may be achieved by using a spraybar as a return tube, a venturi system fitted to the outlet, or even an airstone placed at the point of the returning water.

Coping with power failures

Some thought should also be given as to what happens should the electricity supply be cut off or debris block the inlet of the fluidized bed unit. The water flow through the unit will stop, but, more importantly, the "filter bed" will physically collapse from its previous "in suspension" form to a solid mass of sand. On the resumption of power, the pump may not have enough energy to "resuspend" the sand without a little help, with consequent damage to the pump. (If possible, adjust the flow to maximum on restarting, and then reset it to normal flow rate.) Equally problematical for the aquarium would be the rapid demise of the bacterial colony should the power failure last an appreciable length of time. When restarting after a lengthy period, run the unit in a monitored "closed system" situation until tests show the bacterial bed has become reestablished.

Refugium

A refugium is a body of water, separate from the main tank but sharing its water source. On one hand, it can live up to its name and be nothing more than an area of shelter. Alternatively, and more likely, the refugium can be used in several ways in conjunction with the main aquarium. It provides ideal growth conditions for food items, such as copepods and amphipods. Depending on design and installation, these animals can be continuously swept back into the main tank to feed both invertebrates and fish. It can also be a controlled growing area for macro-algae. Caulerpa and other macro-algae will not only consume ammonia/nitrates and phosphates that accumulate due to overstocking or overfeeding but, because they are grown in isolation, they avoid being preyed upon by vegetarian-minded fish. Other uses include acting as a convenient parking area for heaters, UV sterilizers, etc., a temporary isolation area for a bullying species or as a breeding tank. It has been suggested that daily pH fluctuations can be avoided by running the refugium in a reverse photoperiod operation. A further adaptation uses a special substrate that provides a continuing supply of vital trace elements to the main tank and also harbors beneficial anaerobic denitrifying bacteria.

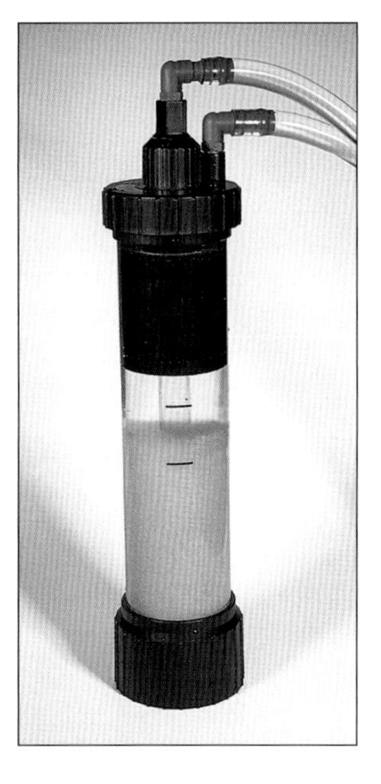

Above: *A typical fluidized bed filter in operation. The sand's moving biological bed is held in suspension (the two lines show the upper and lower levels of the medium). A valve in the unit prevents sand siphoning back into the pump if the power fails.*

A fluidized bed filter

Control the rate of water flowing into the filter by rotating this cap.

Water from the aquarium

Water returning to the aquarium

This cartridge is filled with activated carbon to adsorb impurities from the water.

The media used in the filter is a special silica sand. Once in suspension, it forms a "biological bed" with an enormous surface area.

An optional cartridge can be fitted to remove phosphates from the water.

Ultraviolet light sterilization

Ultraviolet (UV) light is capable of eliminating algae spores and bacteria that are freely suspended in the water and sometimes cause cloudiness. A germicidal-grade lamp can also destroy disease pathogens but is only effective on pathogens with a free-swimming stage that may find their way through the UV unit, so do not rely on it for total disease control. Bear in mind that UV light is dangerous to the eyes: NEVER look directly at the lamp without eye protection.

The UV sterilizer consists of a UV lamp fitted inside a sealed, clear, quartz sleeve, mounted within a water jacket. Aquarium water fed into the outer jacket is sterilized as it flows from one end to another. The efficiency of sterilization depends, among other factors, on the length of time the water is exposed to the UV light. The effectiveness of the lamp can be optimized by prefiltering the water mechanically before passing it through the sterilizer. This prevents organic material and suspended matter from obscuring the quartz sleeve and thus lessening the effect of the UV rays. Even so, the lamp has a definite useful lifespan; used continually, most lamps will require replacement every six months. After this period, the lamp may still appear to be functioning, but in fact the UV light is no longer able to penetrate the quartz sleeve and sterilize the water.

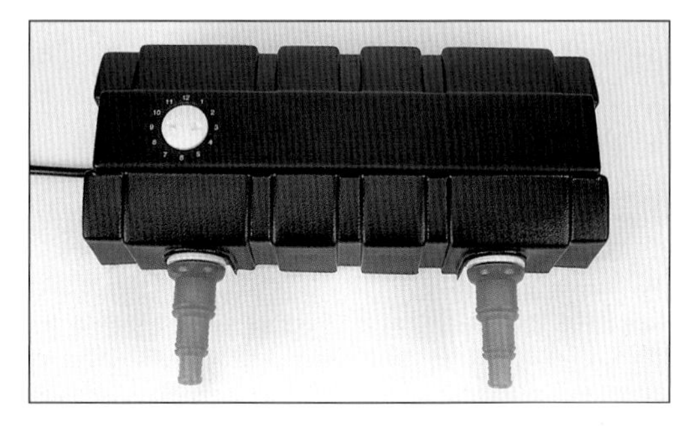

Above: *This is a typical UV unit used to treat aquarium water. The water hose connectors are translucent so that you can see the "glow" of the UV tube when in operation. This is a safety measure to make sure you do not open the unit and damage your eyesight by looking directly at the lamp.*

How a UV sterilizer works

High-energy UV light disrupts living cells in a number of ways. The genetic material (DNA) in the nucleus and tiny "power packs" called mitochondria dotted around the cell are all damaged by the rays. In algae cells, the light ruptures the chlorophyll-containing plastids that carry out essential photosynthesis.

The fluorescent tube in the middle produces UV light with a wavelength of 253.7 nm. This UVC is harmful to living tissue.

This quartz sleeve encases the tube but allows UV light to pass through.

Water flows though this outer glass tube.

Above: *Aquatic dealers run several filter systems to sustain their livestock, including heavy-duty UV sterilizers such as these.*

Using ozone

Ozone (O_3) is an unstable form of oxygen (O_2). The extra atom of oxygen readily separates from the molecule and oxidizes toxins and other compounds in the aquarium water. This oxidizing action makes ozone an effective disinfectant; it will kill bacteria, disease pathogens, algae spores, and other free-swimming microorganisms that come into close contact with the gas. Ozone is used by many large public aquariums as part of their overall water quality management systems.

Ozone is formed by passing air (ideally dried) over an electrical discharge in a device known as an ozonizer, converting some of the oxygen in the air to ozone. The ozonized air is then passed through a special water/ozone contact chamber or a protein skimmer. Providing this is made from ozone-resistant materials, it makes an ideal ozone reactor. Since ozone is harmful to the aquarium inhabitants, it must only be applied in a separate vessel, never directly into the aquarium. Ozone-treated water should be passed through a carbon filter before being returned to the aquarium in order to remove any residual ozone or any harmful by-products of excessive ozonization. It is advisable to use an ozonizer together with a redox (or ORP – oxidation, reduction potential) controller. A probe in the aquarium monitors ozone concentration in terms of the redox potential of the water and switches on the ozonizer only when the water conditions require it.

As you might imagine, ozone is a dangerous gas and must be used with the utmost care. Always follow the manufacturer's instructions to the letter. Ozone is also capable of damaging materials, such as rubber piping and pump diaphragms, and excessive levels can also make plastics turn brittle.

Left: *With their crowns of feathery tentacles, tubeworms make attractive subjects for the marine aquarium. They will thrive in strong currents of clean water but need fine forms of food to sustain them.*

Ozone and protein skimming

Above: *Activated carbon placed in a compartment on top of the collecting cup prevents any excess ozone entering the atmosphere.*

Froth carrying organic waste overflows into this collecting cup.

The waste can be drained from the base of the collecting cup through this tube.

The ozone generator uses a high-energy electrical discharge to bond an extra atom to oxygen molecules to create ozone (O_3).

A check valve prevents water siphoning back into the ozonizer. Renew this regularly.

Ozone passes into the water flow through the venturi device located here.

This probe hangs in the aquarium. It measures the redox potential of the water and regulates the amount of ozone produced by the ozonizer.

The Jaubert system

Until relatively recently, marine aquariums relied on mechanical, chemical, and biological methods of filtration by using long-established conventional units. One innovative progression in the 1980s was the Jaubert system, best described as a subgravel system without any uplift tube or any deliberately engineered water flow through it. Primarily designed for use in Dr. Jaubert's own Monaco Oceanographic Aquarium, this system's basics soon attracted interest and several American marine fishkeepers, such as Bob Goemans, adapted it with some success for use in the more modest-sized home aquarium.

A plastic grid (fluorescent light diffusing grids are ideal) is held off the bottom of the tank by plastic supports to create a static void (plenum). A thin layer of synthetic floss material covers the grid to prevent material trickling downward and the rest of the substrate placed over it. The natural diffusion of oxygen downward from the main, additionally aerated body of water toward the oxygen-depleted plenum sustains the biological filtration action within the substrate, while bacteria in the anaerobic plenum act as denitrifiers, reducing nitrate levels, and completing the full nitrogen cycle.

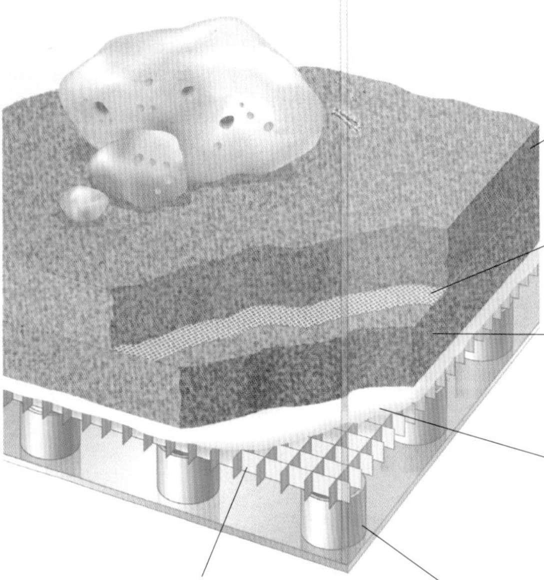

The top layer of substrate – about 1.5 in (4 cm) deep – is aerobic because of the oxygenated water above it.

A layer of plastic mesh prevents disturbance of the substrate by digging or burrowing fish.

The lower level of substrate, and the water-filled void beneath, are anaerobic due to the lack of water movement through the substrate.

A layer of filter floss prevents substrate trickling through the "egg-crate" grid.

Grid formed of criss-crossed plastic slats.

Short pillars support the grid to create a void about 1 in (2.5 cm) deep beneath the substrate.

Using living rock in the aquarium

While a basic aquarium has a protein skimmer and an external filter, the "biological" purification of the water can also be achieved by using another natural method making use of "living rock." This is simply rock collected from the wild with living organisms already in situ. The base material is usually of a labyrinthine, highly porous nature and provides both aerobic and anaerobic areas inhabited by the various minute animals and bacteria. These actively purify the water in much the same way as the bacteria in an external filter. This system has become known as the "Berlin method."

In the Berlin method, more reliance is placed upon living rock and protein skimming in the aquarium to rid it effectively of unwanted toxic materials. The theory is that the protein skimmer removes organic matter before it degenerates into ammonia and therefore obviates the need for its subsequent conversion to nitrite and nitrate by bacteria. The inclusion of large areas of live rock also reduces the load on biological filtration systems. One drawback of using live rock is that it is expensive and furnishing the aquarium with a suitable amount may prove to be a financial deterrent to many fishkeepers.

Live rock is maintained at the aquatic store and may even be inoculated with new living cultures during this time. Obviously, taking the usual pre-cleaning precautions as described for ordinary "decorative" rocks would not apply in this instance. It is vital that living rocks are introduced into oxygenated water as soon as possible. Depending on their quality (a known history is highly preferable), you may put them into your main collection immediately, although treating them to a quarantine period (as you would with any new additional livestock) could be a prudent measure. During quarantine, it may be likely that some organisms will die off and new ones reemerge. Fluctuations in nitrite and nitrate levels will occur, and the conditions should be left to stabilize before transferring the rocks to their new permanent home.

Unless the aquarium is to be set up with living rock from the outset (as shown in the setting up sequence starting on page 46), only small amounts should be introduced into an established system at a time to avoid fluctuations in redox potential.

Reverse osmosis

Thus far we have only discussed equipment required for treating water once it is in the aquarium. However, one of the easiest ways to get ahead in the quest for good water quality is to ensure that only the very best water is used in the first place. To make it safe for human consumption, tapwater contains not only chlorine and chloramine but also pesticides, nitrates, phosphates, and silica, all of which are detrimental to the health of marine animals. To prevent these substances from ever entering the aquarium, you are strongly advised to use a reverse osmosis (R.O.) unit.

Reverse osmosis is a process that removes salts and minerals from the water, plus the toxins mentioned above, as well as some bacteria, viruses, and fungal spores. The key to an R.O. unit is a partially permeable membrane that allows water to pass through but repels all the other substances. Typically, an R.O. unit will have a carbon prefilter to remove chlorine from the incoming water.

An R.O. unit is usually simple to plumb into a domestic coldwater supply. The only real drawback is that typically three to four times as much "reject" water has to be discarded for every measure of "good" water. However, the discarded water is suitable for watering plants, as it is high in useful nutrients.

Using a good-quality salt mix will ensure that the correct trace elements and minerals are restored to the R.O. water before it is added to the tank. If you are using R.O. water to replenish freshwater, you must use a marine buffer and trace element additive in conjunction with the water change.

Above: *Tapwater is processed for human consumption. Chemicals added to it by the local water company make it safe for us to drink but dangerous for aquarium use without prior treatment.*

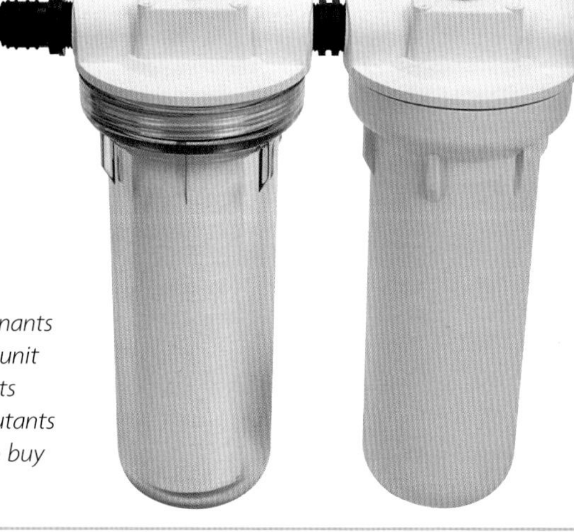

Right: *Tapwater can contain contaminants that can be removed by using an R.O. unit such as this. The efficiency of such units depends on the concentration of pollutants present in the water supply. Be sure to buy the correct unit for your usage.*

How reverse osmosis works

When separated by a partially permeable membrane, the natural tendency is for solutions to equalize in concentration by osmosis. Applying pressure to reverse this process creates very pure water.

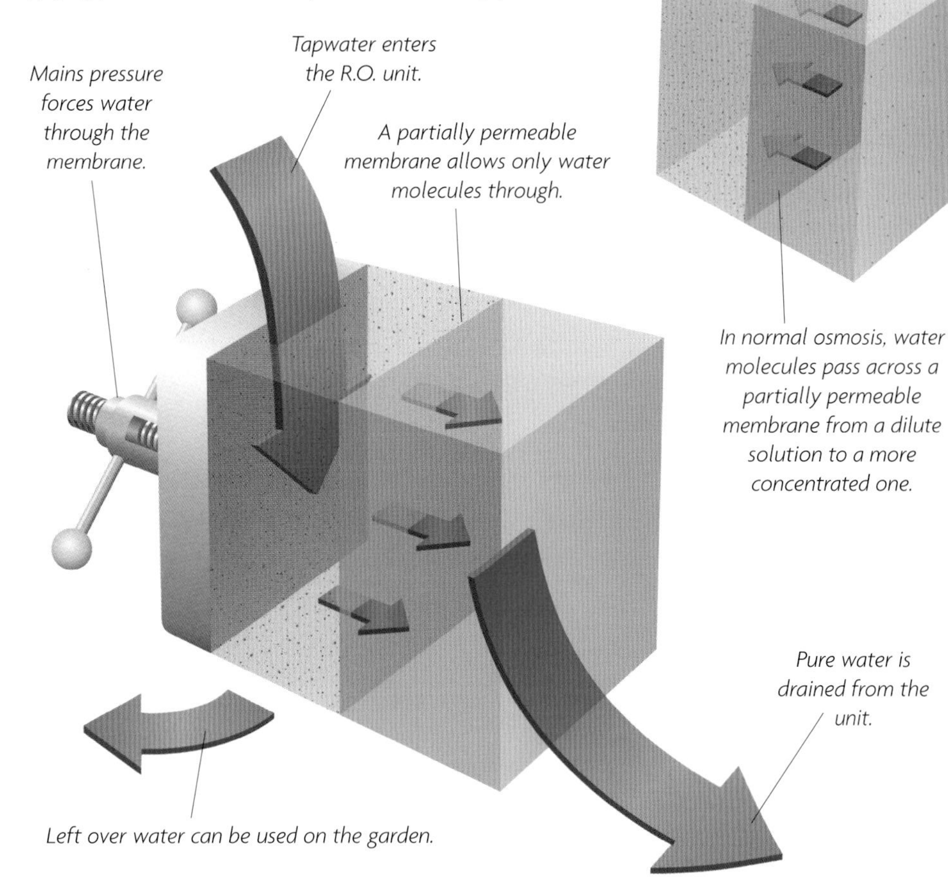

Complete water management systems

The sump beneath a reef aquarium is a "high-tech" environment, full of tubes and boxes that maintain the water conditions in the display tank above it. Water reaches the sump by overflowing from the main aquarium and, once cleaned, is returned by a pump situated in the sump. In the "wet-and-dry" filter, the incoming aquarium water is fed over a mass of plastic balls on its way to the other water treatment components. Being in the atmosphere rather than totally submerged, the nitrifying bacteria that flourish here are well provided with oxygen and so operate more efficiently.

Nitrates can be removed by means of a denitrator unit in which bacteria live in anaerobic conditions. As they "feed," they take their necessary oxygen from the nitrates, thus reducing them back to nitrogen gas.

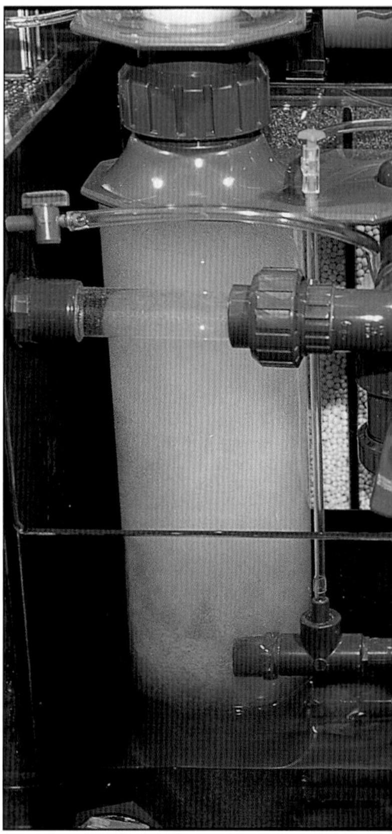

Left: *Complete systems house filtration and other water management devices, including protein skimmers, phosphate removers, freshwater replenishment reservoirs, and even the heating system.*

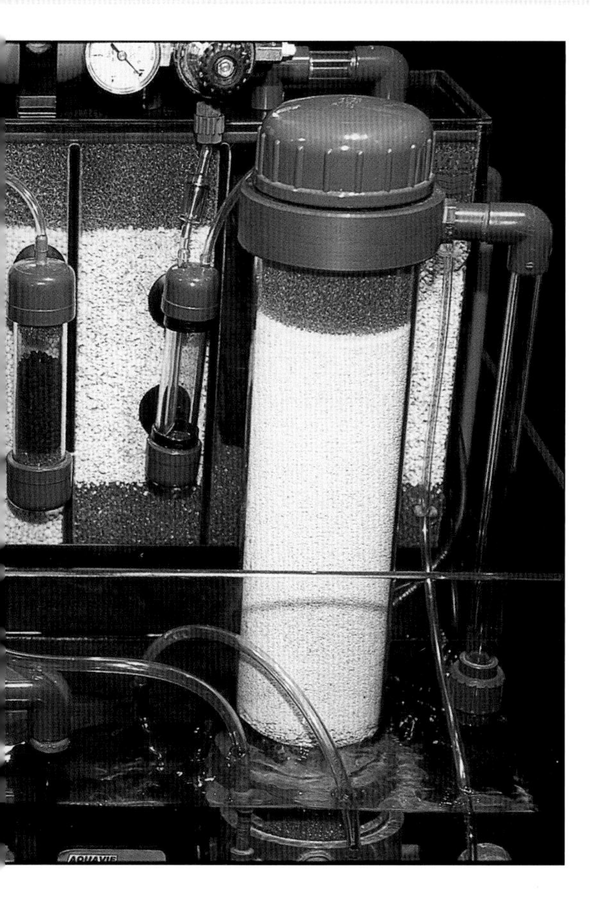

Above: *Where the aquarium is housed in a cabinet, it makes good sense to have the water treatment equipment assembled in one convenient place - out of sight but easily accessible for maintenance tasks.*

Calcium reactors

A calcium reactor is not a filter, but it is included here because it is normally used in tandem with a filtration system. Invertebrates need calcium carbonate in order to grow their skeletons or shells, and regularly take these elements out of the water. As aquarium water ages and its calcium buffering capabilities diminish, the pH has a tendency to drop. This can cause metabolic problems for fishes and invertebrates, such as respiration difficulties. It is vital to maintain the pH at 8.3 for reef animals. A calcium reactor can help to achieve this. The apparatus is usually a sealed acrylic container filled with calcareous material such as aragonite. (Shown here empty.) Aquarium water is delivered to the unit fairly slowly via a bypass system from the main filter. The system is injected with carbon dioxide gas and its acidic nature slowly dissolves the more alkaline aragonite, thus charging the aquarium water with calcium. The carbon dioxide gas is used up in the process.

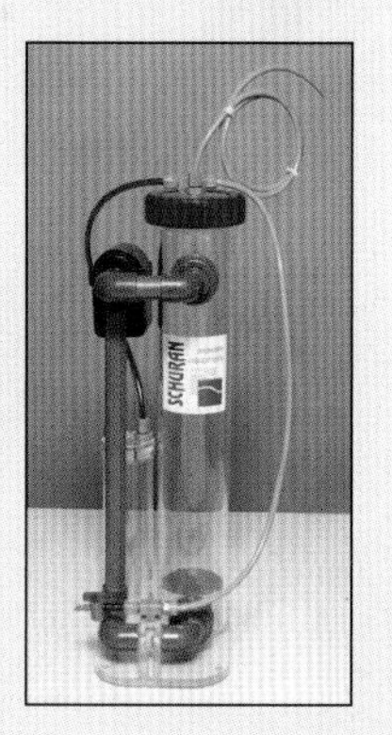

Temperature control

The marine aquarium must be kept at a reasonably constant temperature if fishes and invertebrates from tropical coral reefs are to survive. In some places, measures to cool down the tropical aquarium may be required, particularly during summer months.

Heating equipment may also be incorporated within the water treatment compartment (or sump) of a built-in "total system." This can make more efficient use of your heater, as it is heating a smaller body of water at any one time.

Types of heating equipment

The most popular method of heating aquarium water is by means of small, individual electric immersion heaters combined with a thermostat control, commonly known as heaterstats. These units are generally protected inside a glass sleeve, although you can also buy more rugged types made from materials such as titanium. Some external, canister filters also incorporate a heating system, but be sure to check these out carefully as most are only safe for use in freshwater systems because the elements and thermostat connections could corrode in saltwater. Undergravel heating cables controlled by an external thermostat are also available.

Controlling heat

In a standard aquarium heaterstat, the built-in thermostat controls the supply of electricity to the heating element by sensing the temperature of the water. This sensing is achieved by means of a bimetallic strip that bends and straightens, making and breaking contact as the temperature changes, or by microchip circuitry. Microchip thermostats are normally a separate unit from the heating element and temperature control is achieved by a probe in the water that sends a message to the thermostat, which regulates the power to the heating element.

For most tropical marine aquariums, the ideal temperature is around 75°F (24°C). The familiar heaterstats are usually factory preset to this temperature but can be adjusted if necessary.

Heaterstats are available calibrated in °F or °C, or both scales side by side.

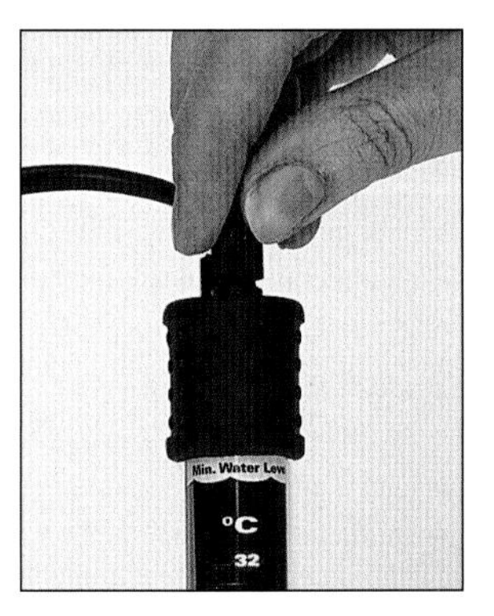

Right: *Heaterstats are easy to adjust by turning the knob at the top until you reach the desired temperature. Some units have a light to indicate whether it is on or off. Make sure you can see it. Always keep a spare heaterstat handy in case of failures.*

External thermostat units can be either analog or digital. Analog units are set manually, but digital models allow you to program a variety of temperature parameters and often have automatic features, such as emitting an alarm if the temperature gets too high or too low. Some include a display that records the maximum and minimum temperature reached in a time period, as well as showing the current temperature. Such digital units can offer the degree of control and accuracy that will provide extra protection for aquarium subjects and peace of mind for you, the aquarist.

Computerized aquarium systems that can control temperature along with a variety of water quality testing features from a single unit are also available. Such equipment often includes software and connecting cables to link with a personal computer, allowing you to plot trends and store other information that you will find useful.

What size heaterstat?

Allow about 10 watts per 1 gallon (4.5 liters) of water content. An aquarium measuring 36x15x12 in (90x30x38 cm) has an approximate volume of 24 gallons (110 liters) and would therefore require 240 watts of heat. A smaller-sized heater would be sufficient if the room is permanently heated. Following the guidelines described above, a single 250-watt heater should suffice, but two separate 150-watt heaters would be better and afford some margin for error or in the event of heater failure.

Keeping it cool

Chillers contain a compact refrigeration unit that cools the incoming water to a preset temperature. Always choose one designed specifically for use with marine aquariums. Ask your dealer for advice on the correct size and type of chiller to use.

Above: *The in and outflow pipe connectors are visible at the front of this chiller.*

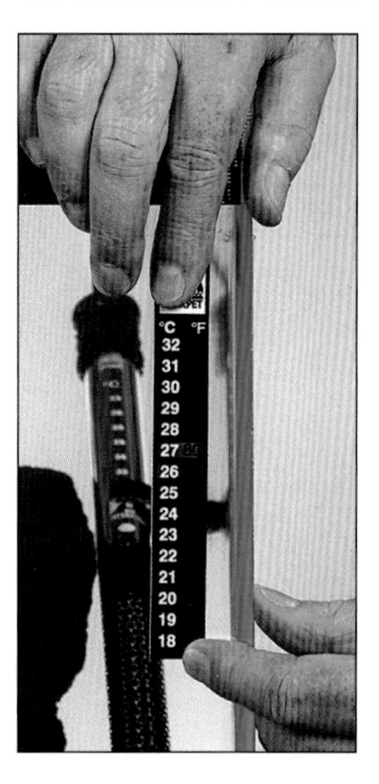

Right: *Various thermometers are available for aquarium use, including easy-to-read liquid crystal types that stick to the outside surface of the tank. A digital model is shown being used on page 57.*

Lighting the aquarium

Lighting in a fish-only aquarium is really down to personal taste. Two fluorescent tubes will usually suffice, with perhaps a third, blue lamp for night viewing. However, if there are invertebrates in the tank, lighting becomes far more important. Many corals, anemones, and mollusks rely on algae known as zooxanthellae for food. As well as good water quality, the main requirement of zooxanthellae and macro algae, such as caulerpa, is high-intensity light at the correct wavelength. In fish-only systems with relatively low light intensities, zooxanthellae will not grow. In a reef aquarium, the correct lighting is therefore of paramount importance, and this usually means using metal-halide lamps.

There are two key elements that contribute to the suitability of a particular lamp for a given marine aquarium setup: light intensity and color temperature. Understanding these factors will help you to choose the best lighting system for a fish-only tank or a reef aquarium with invertebrates.

Light intensity

The electrical power of a lamp is measured in watts, whereas the intensity, or brightness, of a light source is quoted in lumens. (You may also see light intensity quoted in terms of lux – lumens per square meter.)

The intensity of a lamp is usually indicated on the packaging. Most fluorescent tubes rarely exceed 5,000-6,000 lumens, and some are far lower. On the other hand, metal-halide lamps commonly emit in excess of 10,000 lumens.

Light intensity also depends on the efficiency of the lamp and it is difficult to achieve high levels of illumination with fluorescent tubes. This is especially true in deeper tanks, where tube light fades quickly.

Color temperature

Any type of lamp appears to give off light in a particular color. In fact, the color we perceive is made up of many different colors, each with their own spectral wavelengths. This is called the "color temperature" of the lamp and is measured in degrees Kelvin (K). The higher the Kelvin rating, the more white and "cool" the light appears. A low Kelvin rating produces a "warm" appearance, usually mainly red or yellow. For example, a candle flame has a color temperature of 1,800K, a cool white fluorescent tube is 4,000K and a very bright metal-halide lamp can be 14,000K. By contrast, looking up into a clear blue sky over a tropical reef can register up to 30,000K.

The real colors of white light

A rainbow reminds us that white light is made up of a spectrum of colors. They are produced by different wavelengths; violet is short, red is long.

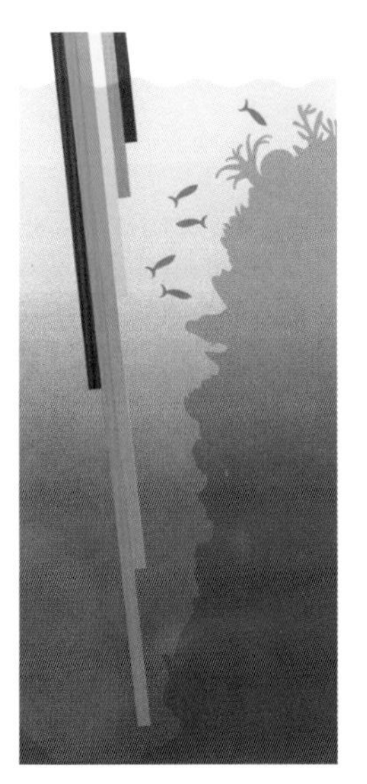

Right: *Various wavelengths of light penetrate to different depths in the sea. Blue light can reach depths of over 820 ft (250 m), while red light fades at about 33 ft (10 m). Ultraviolet light (invisible to us) can reach 330 ft (100 m).*

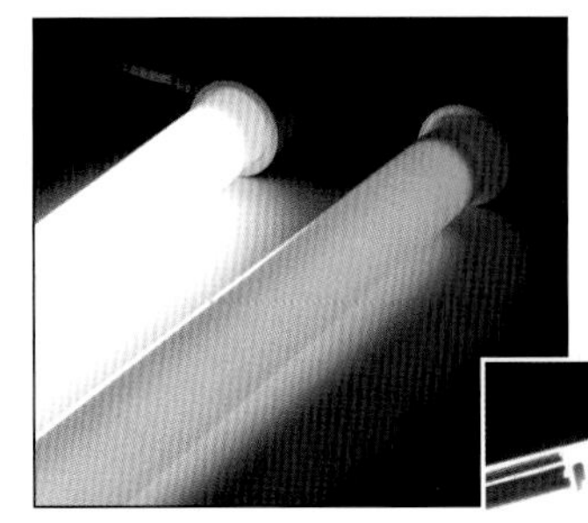

Below: These new forms of the standard white and blue tubes produce 80% more light. Reflectors boost the light even more.

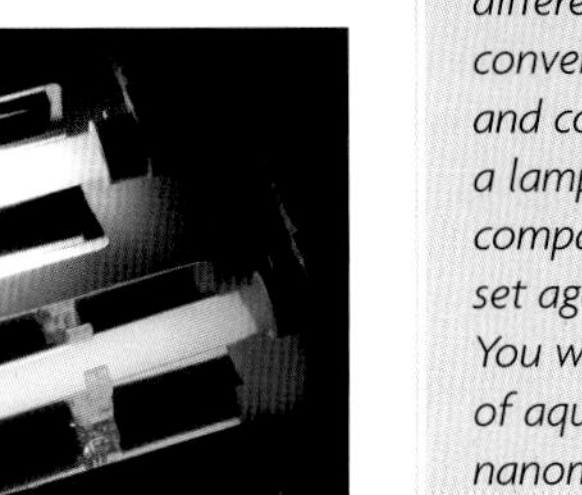

Above: These standard marine white and actinic blue tubes are 24 in (60 cm) long and consume 18 watts of power.

Spectral power distribution

The output of the various wavelengths of light produced by a lamp is referred to as its spectral power distribution. In effect, this represents the strength of light in the different color "bands" and forms a convenient way of graphically illustrating and comparing the color performance of a lamp. The graphs shown below compare the output of two light sources set against the sun's complete spectrum. You will see similar graphs on packaging of aquarium lamps. Wavelengths are in nanometers (nm) – billionths of a meter.

White triphosphor fluorescent tube

These spectral output curves show that this type of fluorescent tube produces useful levels of light at a wide range of wavelengths, providing bright illumination for all creatures in the aquarium. (The vertical scale on these graphs reflects comparative output levels.)

Blue actinic 03 fluorescent tube

This tube is "strong" in the blue area of the spectrum, especially in the so-called actinic range peaking at 420 nm, which is vital for zooxanthellae to thrive. It also supplies some UV for a fluorescent effect.

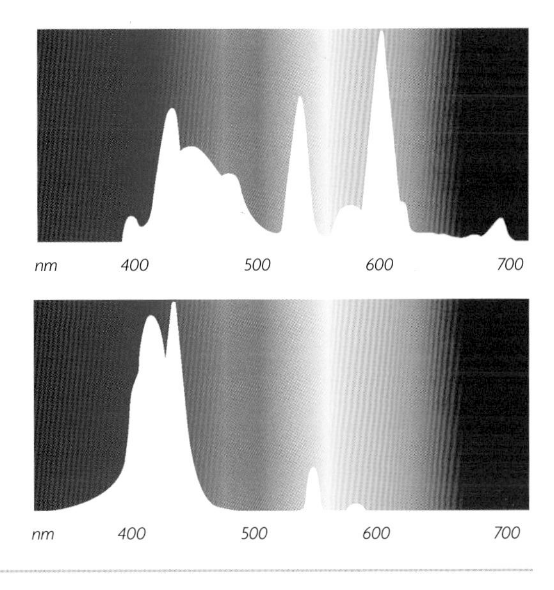

Lighting the aquarium

Left: *The familiar fluorescent tubes used across the world in homes and offices are fine for aquarium use. The development of a triphosphor coating has boosted the output across a wide range of light wavelengths.*

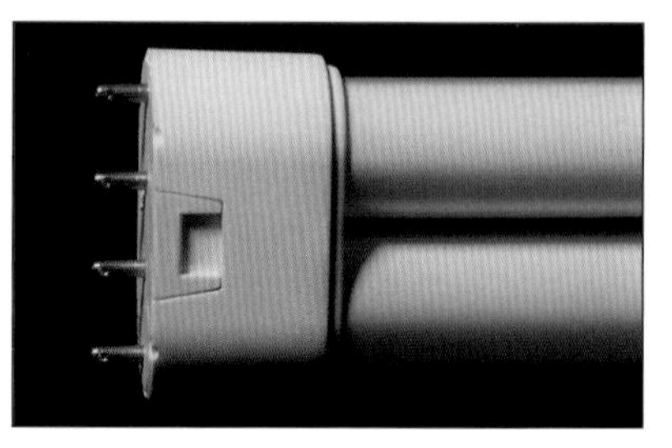

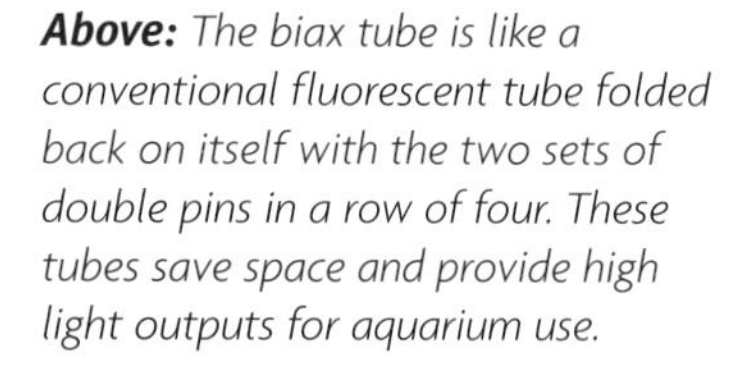

Above: *The biax tube is like a conventional fluorescent tube folded back on itself with the two sets of double pins in a row of four. These tubes save space and provide high light outputs for aquarium use.*

Left: *A narrow-bore form of fluorescent tube provides intense light and saves space in the aquarium canopy. These are triggered with an electronic starter (ballast) system and are available in a range phosphor types.*

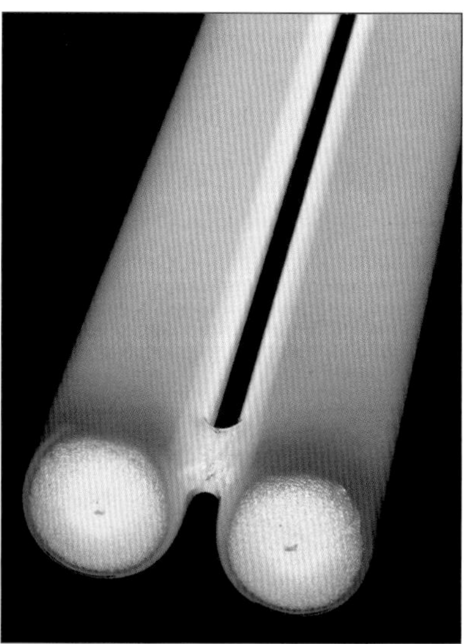

Right: *Powered up, a biax tube glows just like a conventional fluorescent tube. The twin tubes close together produce an intense light for the space they occupy. Another benefit of these lamps is that the tubes can be coated with different phosphor formulations, making it possible to recreate the effect of two separate (and different) tubes in one fitting.*

Right: *Metal-halide lamps for aquarium use are fitted into housings such as this and are suspended above the open tank. They provide very bright illumination but must have some ventilation space around them because of the heat they generate.*

Above: *The provision of white and blue fluorescent tubes in combination with a metal halide lamp offers the fishkeeper the facility of creating separate day and night lighting effects, together with a "hot" highlighted area to feature specially chosen decor or invertebrate life in the tank.*

Metal-halide lamp

Metal-halide lamps generate a wide output with high levels at 400-480 nm (good for zooxanthellae) and 550 nm (to simulate sunlight). The volume of the curves in the spectral output shown below indicates a bright lamp.

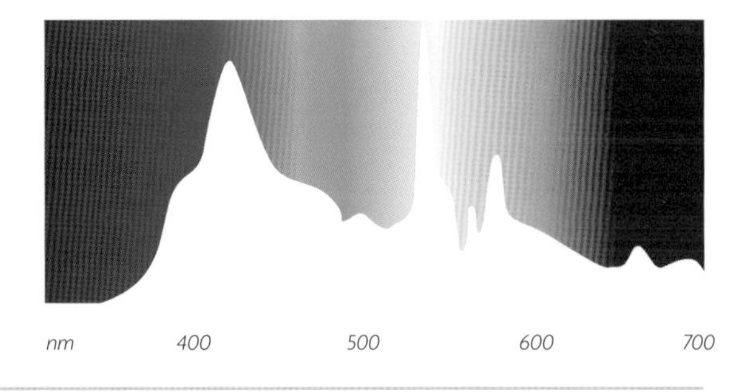

Lighting the aquarium

The final piece in the lighting jigsaw is the length of time the lighting is left on – the "photoperiod." There are no hard and fast rules about this, as every aquarium differs slightly in its requirements. The best way is to learn from nature and try to replicate life on the reef in your aquarium.

Around the equator, the period of sunlight is about 12 hours each day, whereas in tropical latitudes (up to about 20° north and south of the Equator), full sunlight is only experienced between about 9 am to 3 pm (i.e., six hours). As a starting point, if you are using fluorescent tubes, try setting your lighting period to about 12 hours. A blue lamp or one of the less bright lamps should come on one to two hours before the main lighting and go off one to two hours after it has been turned off. This creates a feeling of dusk and dawn and avoids stressing nervous fish.

If you are using metal-halide lamps, try starting with a photoperiod of about eight hours, again with blue lamps that come on before the main lights and remain on after the halides have been turned off for the day. Sophisticated lighting setups include a programmable timer and a special "moonlight" lamp that matches the cycle of the moon!

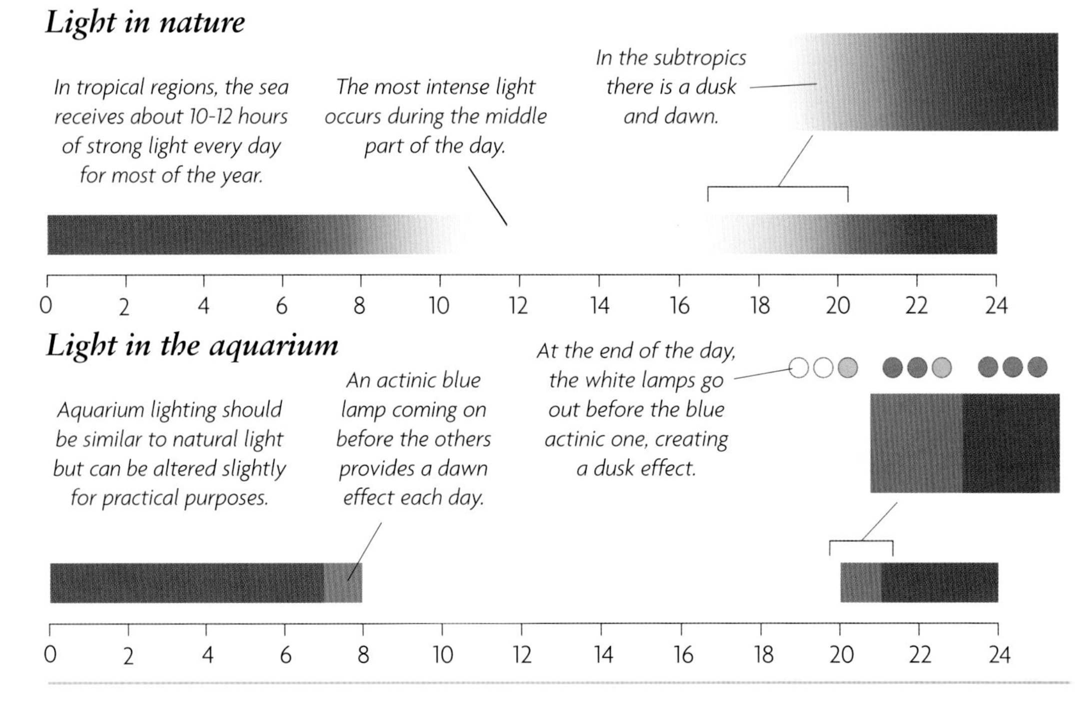

How light is lost in the aquarium

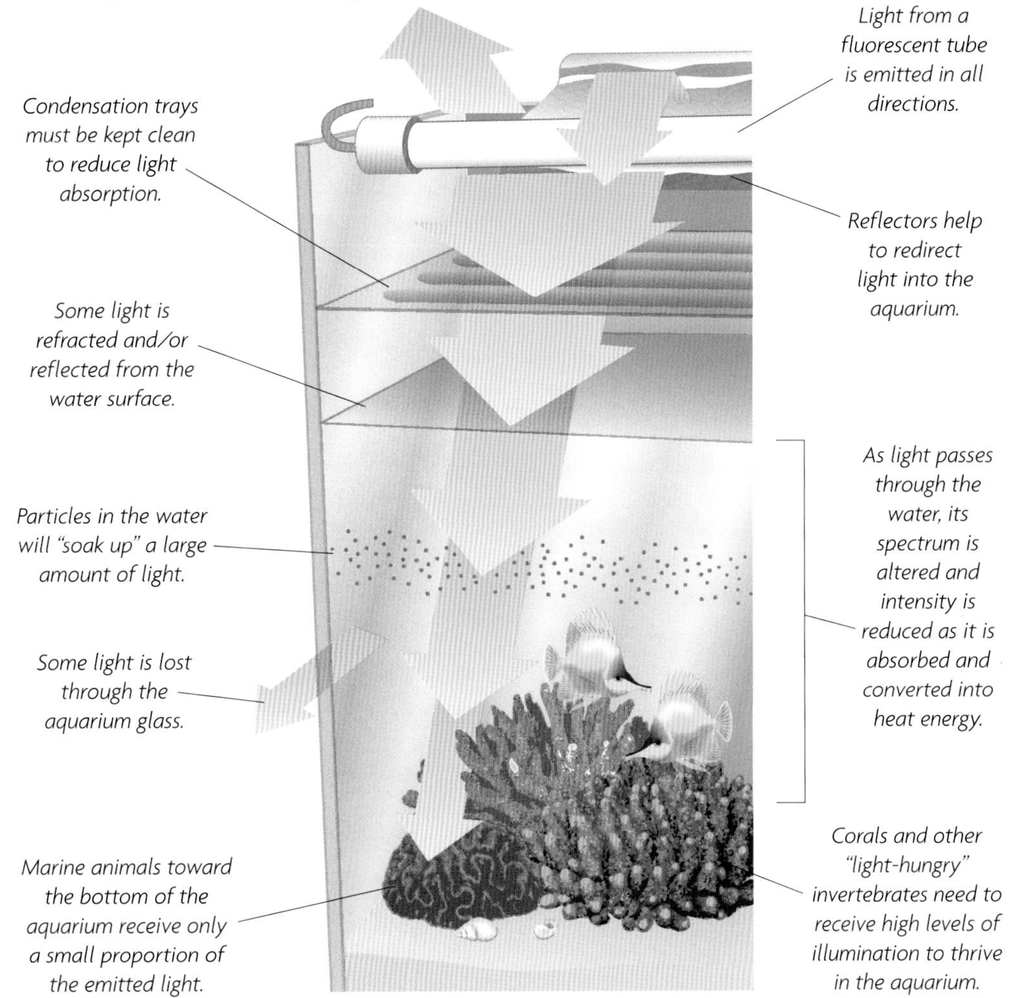

Choosing rocks

Newcomers to marine fishkeeping may be forgiven for thinking that decorating the tank with substrate and rocks, etc., is done solely for aesthetic purposes. However, in modern marine aquariums, these items play an integral role in maintaining water quality and providing the fishes and invertebrates with habitat to live in or on. Generally speaking, these materials require thorough cleaning, or curing, before use and must be prepared in advance of setting up the tank.

When you come to decorate the aquarium, always work from the bottom up. First put in the rockwork or living rock directly on the tank base and then build up the substrate on top. (This is the approach used in the practical sequence of setting up a basic marine aquarium, starting on page 46.) An obvious statement perhaps, but if you intend to use biological filters, such as a plenum, that are often situated below the substrate, then you must fit these before any other work is carried out.

The substrate

In the case of a tropical marine aquarium, coral sand and gravel are the only real choices if you wish to establish a "natural" look. In temperate marine systems, you could use soft beach sand, gravel, or small pebbles, or a combination of these materials.

Left: *Coral gravel and sand (shown here) are the most suitable and easily available substrates for a marine aquarium. They not only look right but also help to maintain a stable pH and high buffering capacity.*

Sand on the reef is mostly produced by animals such as boring worms and parrotfishes, which crunch up coral to extract tiny organisms and algae. Their crunching activities are so thorough that they eventually excrete fine sand that falls to the seabed. Coral gravel is created by wave action, whereby the coral is broken and the fragments are slowly worn down. In nature, myriad organisms, such as worms and crustaceans, some microscopic, live in the sand and keep it clean and aerated to a certain depth. In the aquarium, too, these organisms can become very useful when the sand is allowed to become "live."

Choosing rocks

Apart from deciding how to set out the tank, you must choose the correct materials to put into it. Once again, remember the importance of water quality control and choose from the wide range of calcareous and inert substances available. Fortunately, there is no difficulty in obtaining materials such as lava rock, ocean rock, and so-called living rock (see the panel on page 29) for use in the tropical marine aquarium. Where temperate or coldwater marine tanks are concerned, you could use a wider variety, such as sandstone, granite, some slates (gray is safe) and limestone. Whichever you decide on, you must examine the pieces before use to make sure there is no evidence of metallic-ore veins in the rock. Ores have occasionally been found in tufa and volcanic lava rock, in particular.

How you design the habitat in your aquarium is largely a matter of taste, but it is worth taking a bit

Rocks for a marine aquarium

Calcareous oceanic rock will help to maintain pH levels in the aquarium.

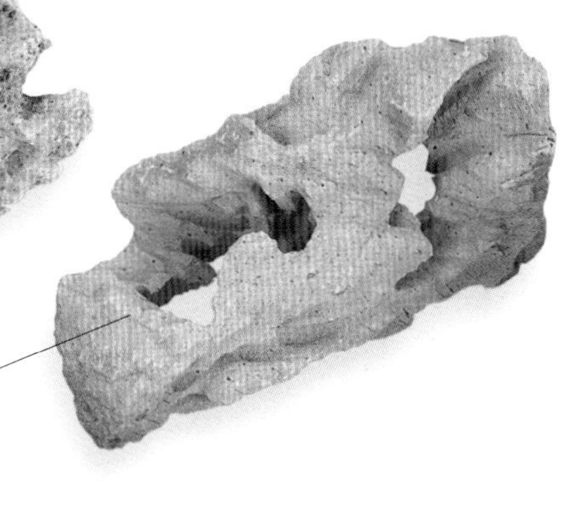

This ocean rock has been smoothed off and will provide an excellent footing for algal growth.

This porous rock has been sculpted by hand to provide "swim-throughs."

of time to consider the needs of the animals you want to keep. Fishes certainly need somewhere to go at night, unless they are nocturnal and the reverse applies. Shrimps and other crustaceans require overhang ledges to get under, and some fishes, such as lionfishes, often hang upside-down under ledges. Territorial animals, as the name implies, must be able to establish their own territories, while cruising species, such as triggerfishes, batfishes, and angelfishes, need a reasonable amount of open space. By including caves, crevices, ledges, and arches, along with plenty of swimming space, you will be able to cater for a wide range of needs.

Building a lava rock cave

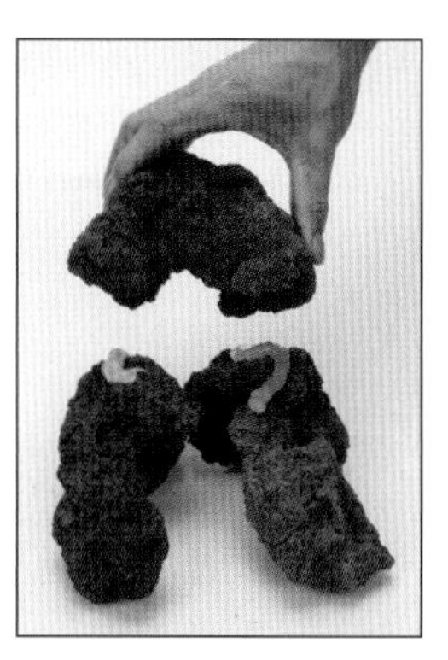

Press the "roof" onto the siliconed areas of two side pieces. Allow to set a few days before use.

Synthetic forms of aquarium decor

Thanks to wildlife TV programs, our ideas about corals for the marine aquarium have not only changed but become more informed over recent years. No longer are the dead skeletal remains of coral considered either decorative or even representative of nature. Nowadays, we appreciate corals for the living creatures they are and go to great lengths to ensure the necessary aquarium conditions are set up almost as much for their benefit as for the fishes.

For those that wish to simulate the undersea world their imaginations provide, without resorting to worrying about keeping real corals alive, there are very realistic replica models of coral heads with which to decorate the aquarium. Closely modeled to represent the real thing, these replicas soon lose their artificiality when covered with a little aquarium detritus or algae growth. They still provide natural-looking refuges and sanctuaries for the fish as they do in nature and have one advantage over live corals in that they can be taken out and given a good scrub should they get too unattractive!

Very natural structures can be built up with artificial rocks by using cable-ties; actually hanging some pieces from cross-struts over the aquarium can avoid the risk of rocks collapsing – especially if there is only a shallow substrate in which to imbed them or if you are trying to create overhangs.

Try to avoid creating dead spots, where there is little or no water flow; it is important that all parts of the aquarium receive well-oxygenated water, particularly in a reef-type aquarium, where water currents will bring food to the sedentary invertebrate life. Judicial placings of small extra powerheads will ensure ideal water movement for this purpose.

Grouping these artificial barnacle shells in a cluster creates a more natural effect.

Left: *Whether the rockwork is natural or synthetic makes no difference to any fish seeking a sanctuary within its many crevices. As darkness falls, most fish will find a hiding place for the night.*

***Right:** By choosing contrasting shapes, sizes, and colors, even these clearly synthetic models of coral heads can add extra dimension to the aquarium's decorations. Bear in mind that these structures will not grow but make sure you have enough room for the live soft and hard corals to do so.*

Although it looks garish out of water, this simulated tree coral will soon lose its harsh appearance and add color and interest to any display.

This synthetic seafan looks surprisingly realistic, and its elegant sweeping form will contrast well with more upright shapes.

***Right:** Models of corals molded in resin are incredibly lifelike and will soon take on the patina of realism as they become covered in algae and detritus in a marine aquarium. They are the only choice if we are to conserve the real thing in the wild.*

Siting and preparing the tank

On the following pages we consider the stages involved in setting up a basic marine aquarium by using separate pieces of equipment to achieve the three main types of filtration – biological, mechanical, and chemical – along with heating and lighting. Our customized aquarium includes substrate, some living rock and a compatible selection of fishes and invertebrates.

Our setup will feature an externally fitted fluidized bed sand filter, an external protein skimmer for removing organic and chemical waste, an external power filter with foam and filter wool to remove particles from suspension and improve water movement in the aquarium. As some of the equipment will be situated outside the tank, make sure there is enough room on the stand (or along the back panel of the tank) to accommodate it.

It is a good idea to compile a list of all the items you will need before beginning the setting up process. Make sure that both the floor and the tank stand or cabinet are strong enough to take the weight of an aquarium full of water. Follow the guidelines for positioning the tank and place a sheet of expanded polystyrene or plastic foam mat on top of the stand to help support the tank evenly. Then sit the tank on top of that.

Cleaning the tank

It pays to clean the aquarium thoroughly before setting it up and you can do this fairly conveniently once it is in its final position. You will find that aquarium accessories attached by "sucker caps" stick much better to clean glass. First remove any packaging tapes, fingermarks, and so on. Wash the inside of the tank with warm, slightly salty water and dry it off with a lint-free cloth or paper towel. A final polish with a cloth soaked in rubbing alcohol will remove any stubborn grease or packaging adhesive. Polish the outside of the glass with commercial glass cleaner if you wish – but never the inside!

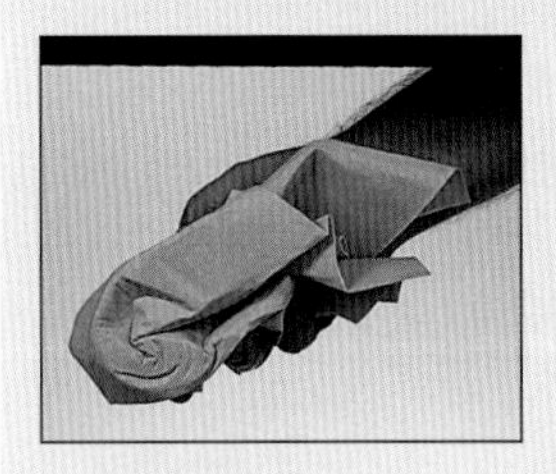

Check the level of the tank in both directions. It must be perfectly level to avoid stresses in the glass panels when it is filled with water. Remember that it is impossible to make leveling adjustments once the tank is full.

Siting the aquarium

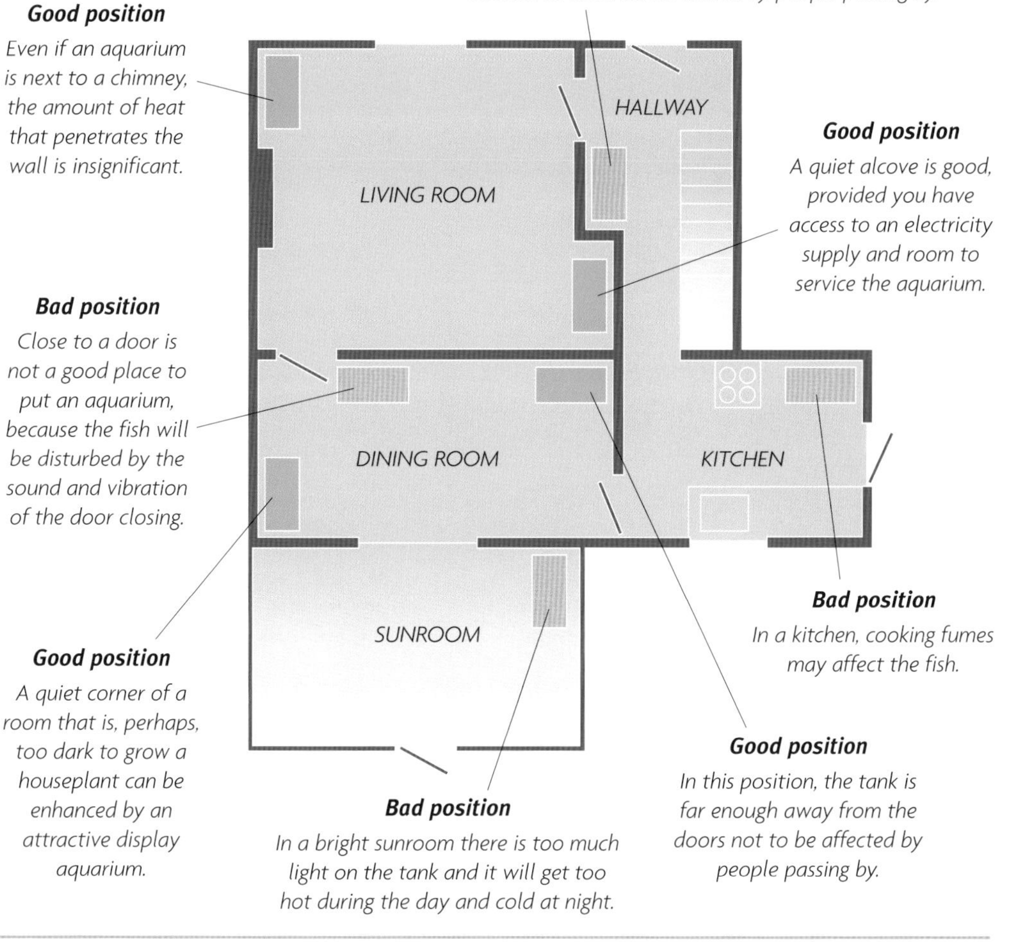

Adding heaters and filtration

Of all the essential technical requirements in a marine aquarium, heating is the easiest to fit and the least demanding to maintain. As explained on page 34, a simple, thermostatically controlled electric immersion heater (heaterstat) hangs in the tank and switches itself on and off as needed. In tanks over 24 in (60 cm) long, it is usual to divide the total wattage of heat required into two, and install two heaterstats, one at each end of the tank. As well as ensuring an even spread of heat, you can be reassured that if one of the heaterstat should fail, the other will act as a backup.

Heaterstats are wired into the electricity supply, usually through a "cable tidy" connecting box. Unlike pumps and lights, they are NOT serviced by switched circuits. When carrying out maintenance within the aquarium, switch off the power supply at the power outlet before putting your hands into the tank. Allow some minutes to elapse following "switch off" for any residual heat in the heaters to expend itself before you work on the tank, especially if it is to be drained.

Fitting heaters is an easy task; simply attach their supporting suction caps to the glass of the aquarium. To ensure long-lasting suction, fix the suckers to the dry glass before adding the water. As water is added, it will increase the pressure on the suckers. Fitting suction caps under water is not so successful, as there is always a small amount of water trapped between the suction cap and glass that minimizes the clinging effect.

Always make sure that heaters are mounted clear of the substrate material to allow maximum circulation of water around the heater.

Heating options

The options when heating the aquarium fall between two choices. You may choose to have the heating equipment either in the aquarium itself or as part of a separate water management sump sited below the aquarium. In this case, it may be prudent to put one heater in the sump and one in the tank, just to be safe.

The temperature scale is clearly marked along the top section of this heaterstat. Remember that this shows the heat setting you choose, not the actual temperature of the water.

Left: *Sometimes fishes, especially those of a nervous disposition, may attempt to shelter too near a heater, with the result that they may suffer burns. Heater guards can be fitted around the heater units to prevent this occurring.*

1 *In all but the smallest of aquariums, it is best to use two heaters placed at opposite ends of the tank to ensure even heat distribution and as a safeguard against heater failure. These units combine a heating element with a small electric water pump to distribute heat.*

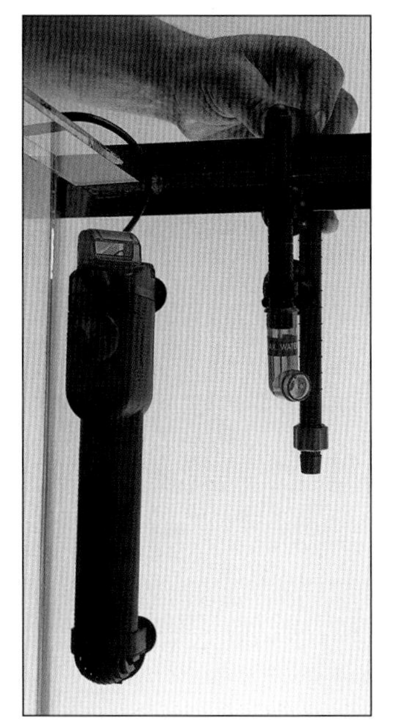

2 *Place the return pipe from the external power filter at one end of the tank and try to arrange it so that it can be easily hidden by your habitat.*

3 *Place the inlet filter pipe at the opposite end of the tank from the return pipe to provide the most efficient water circulation.*

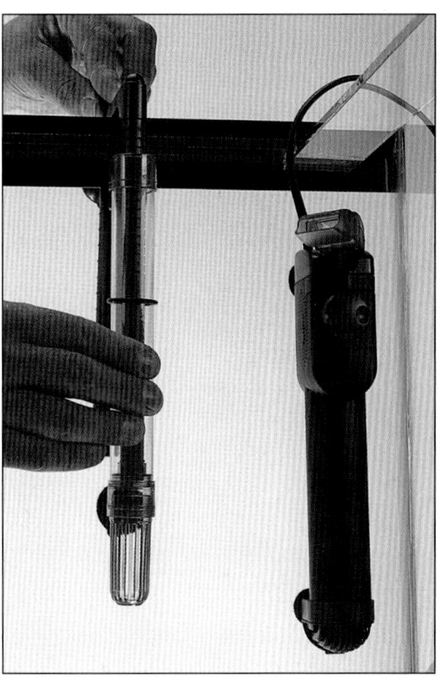

Installing a power filter

If the heater is allowed to lie on the surface of the substrate, local boiling action could occur, with the result that the glass tube containing the heater element may crack. Similarly, do not crowd tank decorations around the heater unit or otherwise restrict water movement around it.

There are other methods of heating the water. Some external canister power filter units have heating elements incorporated into their design. Where the filtration system is made up of a separate sump unit, often housed immediately below the aquarium, heaters may be placed in the sump rather than in the aquarium itself.

To adjust the water temperature, simply turn the control at the top of the heaterstat. A quarter turn is the most you will need in any direction to effect a change. Allow at least 30-60 minutes to elapse before re-reading the water temperature. The action of the heaterstat is generally indicated by a small neon lamp within the heater body; when it is lit, the heater is working.

Do not confuse the heater's "temperature-set" indicator with the actual temperature of the water – always check the thermometer in or on the tank.

4 *Connect hoses to the external power filter. Take care to fit these securely so that they do not come apart when the filter is running under full pressure – otherwise water in the tank ends up on the floor!*

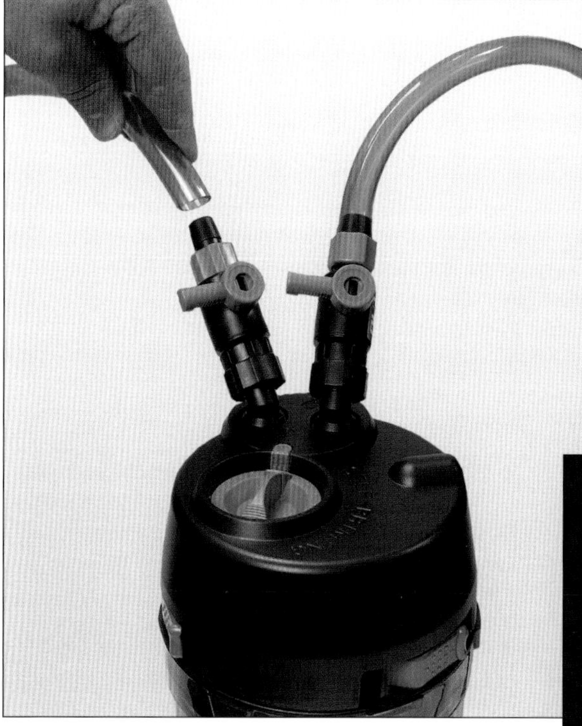

External filter taps

Always fit isolating taps on any external filter. They make filter cleaning simple and mess-free! If you fill the filter body almost full with water after cleaning, there will be very little air to purge from the system once you reconnect and open the taps. Keep a close watch on the water emerging from the return pipe – any slowing down indicates the need for a filter clean.

5 *Connect the hoses to the external filter securely, making sure they are not tangled and, if necessary, mark them IN and OUT.*

6 *If you choose to place your aquarium on a cabinet, then it is easy to hide away external equipment such as this power filter.*

Installing other filtration systems

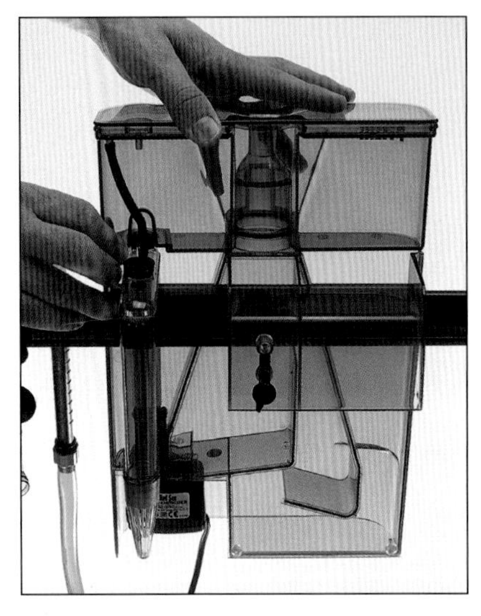

7 There are several designs of protein skimmer available. This one needs to be hung on the back panel of the aquarium. Make sure you provide easy access to such equipment for regular cleaning.

8 This aquarium makes use of a fluidized bed sand filter for biological filtration. This particular unit is driven by a small submersible water pump. Attach this by suckers inside the tank on the back glass.

9 Hang the fluidized bed sand filter on the outside of the back glass and connect it to the internal pump with a hose. Always make sure that the chamber containing the sand is fitted in a truly vertical position.

__10__ Here is how the aquarium looks with the heating and filtration equipment fitted.

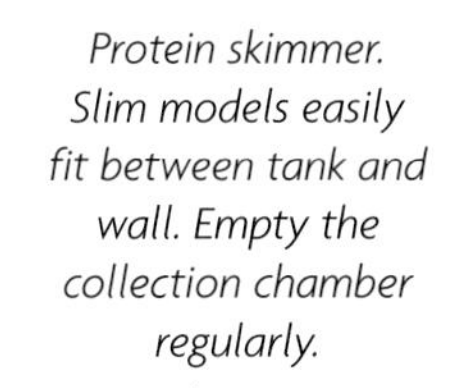

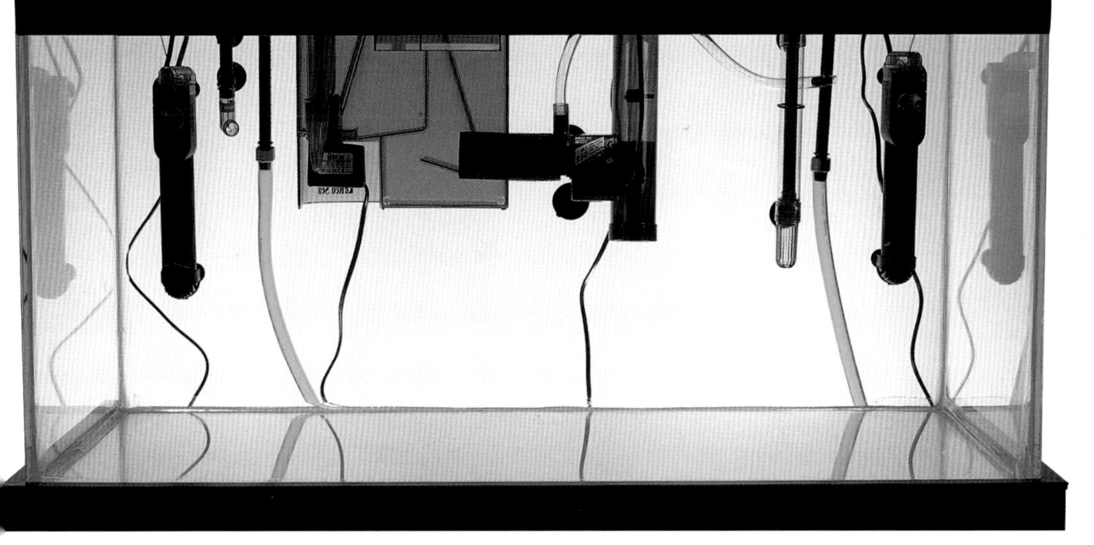

Adding living rock and substrate

Providing you have a supply of previously matured saltwater on hand that has the correct temperature, specific gravity, and pH of 8.3, you can now introduce pieces of living rock. Add the water immediately after placing your rocks and laying the substrate. If you only have newly mixed saltwater available at the setting-up stage, wait until the aquarium is fully operational before adding the living rock to avoid damaging the organisms in it.

In this aquarium we are using a coral sand substrate to a depth of about 1 in (2.5 cm). This is equivalent to about 10 lb (4.5 kg) per 1 ft^2 (900 cm^2); any deeper and you will run the risk of anaerobic conditions developing in the lower layers.

You will probably need to wash the substrate before use, and this is best done in advance of starting the setting-up process. Wash a small amount at a time in a fine mesh hand net. Hold the net over a bucket and run water through from a tap or hose until all the detritus has been flushed out of the sand. Dispose of the waste, but try to avoid clogging the trap in your sink. When you are happy that all the sand is clean, spread it over the base of your tank and around the habitat so that the rocks appear to be rising out of the substrate.

11 *Fit a background to the outside of the back glass to hide the equipment. Blue is the best choice, as the colors of fishes and invertebrates stand out well against it. Black can also be used to good effect in a display aquarium.*

12 *Add the first pieces of living rock directly onto the tank floor. Place them carefully so that you can build up a stable display. In this setup we are using high-quality living rock only and have opted not to include any form of base rock or other aquarium decor.*

13 *Create the habitat in a way to provide plenty of retreats for the inhabitants and to hide pieces of equipment placed internally. However, leave open swimming space for those fishes that require it. It is vital that the rock is not allowed to dry out.*

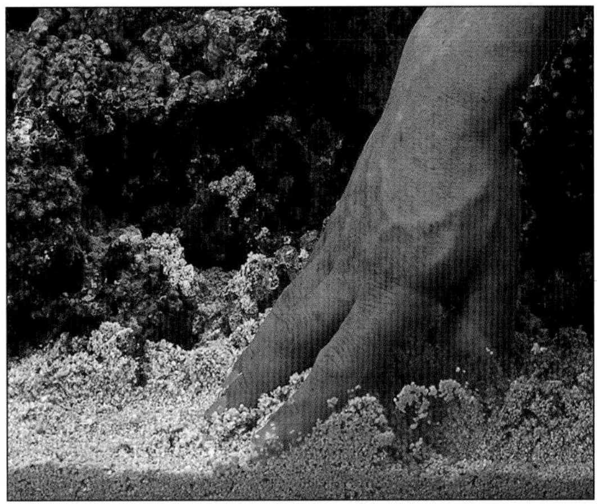

14 *The best substrate for this aquarium setup is coral sand. Spread it around the base of the tank and against the rocks to give the appearance of the habitat rising up from within the sand.*

Adding the water

If you have already mixed the water in another container or spare aquarium, then it is simply a case of pouring it carefully into the tank. It is a good idea to pour the water onto a rock or into a container placed on the substrate to avoid the sand being moved out of place. Now connect the pumps for the fluidized bed sand filter and protein skimmer to the electricity supply and switch on in order to fill their chambers and start circulation. Carry this out in full accordance with the manufacturer's instructions. The water level in the aquarium will drop a little and will need to be topped off. Switch on the power filter and heater.

If you do not have prematured water ready you will have to mix a new batch before setting up. When mixing, the temperature should be at the normal aquarium operating level of 75°F (24°C), as the salts will dissolve more readily. A thermometer stuck to the outside of the mixing container is good as a quick reference, but, to be accurate, use a glass thermometer that reads directly from the water. Once you are certain the salts have completely dissolved take a specific gravity reading by using a hydrometer. If it is low, add more salts and continue doing this while taking further S.G. tests, until the S.G. is 1.022-1.023.

Remember, once the aquarium is fully operational and has animals living in it, you must never mix saltwater directly in the tank again. Always perform this task in a separate container.

When you are happy that everything is correct, add the water to the aquarium and connect up the heating and filtration systems.

Left: *Always use nonmetallic containers in which to mix the marine salts with water. Ideally, food-quality plastic buckets are the best. Look out for nitrate and phosphate-free salt mixes.*

Keep a suitable bucket solely for aquarium use.

Mixing techniques

Modern salt mixes are carefully researched and accurately contain everything to replicate "seawater" for the aquarium. You should use a spare aquarium heaterstat in the mixing container to ensure that the mix is at the correct temperature before you check the specific gravity. Aerate the mix strongly for 24 hours before using.

15 *Because this setup includes living rock, it is important to add matured saltwater, i.e., from another established aquarium. New saltwater can harm the many tiny invertebrates already living on the rock. Pour the water gently onto the rocks to prevent displacing the substrate. Keep living rock wet until ready for use in the tank.*

16 *Attach the readout unit of a digital thermometer to the outer glass and place the probe in the water, ideally some distance away from the heating units.*

Adding lights and maturing the tank

The lighting canopy must be able to accommodate two white 10,000K fluorescent tubes and one blue actinic 03. It is also important that it can have sections cut out of it to allow filter fittings, hoses, and cables to be routed through. A wooden hood laminated with plastic, or painted with several coats of polyurethane varnish, is ideal. Site any control gear for the lamps away from water and splashes.

It is worth connecting the power cables for the lights into a remote control box or power strip with the pumps and heaters so that there is only one cable running to the electricity supply socket.

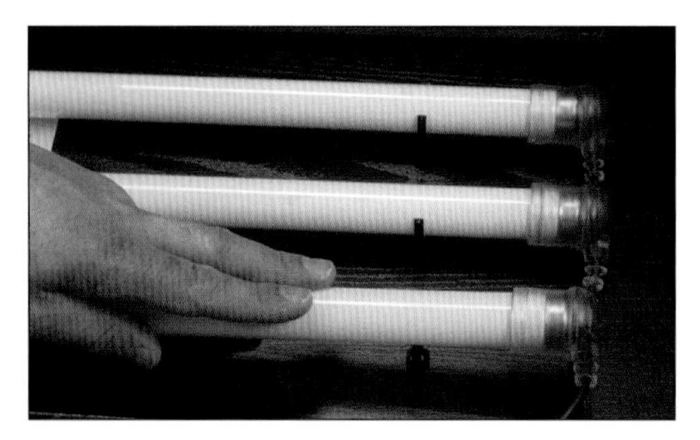

17 Position the tubes in the hood (blue at the back) so they provide the best coverage within the aquarium. Place each tube as far apart as possible from its neighbor to allow for good air circulation between them, which prevents overheating.

Maturing the filters

Even though there are no animals in the aquarium at this stage, leave the lights on for the normal photoperiod, i.e., 12 hours per day, as this will help the maturation process. It is essential once the living rock has been introduced.

It is absolutely critical that you do not introduce any more livestock into the aquarium until the filters have matured. This means you must wait until there is a large enough population of nitrifying bacteria to deal with ammonia and nitrite. As the "living rock" in this setup may have bacteria already living on it, the process could be faster than in other arrangements. Make daily tests for ammonia and nitrite levels until the readings show zero for both. It is likely you will not see any evidence of ammonia or nitrite right away, but this does not mean the coast is clear, as the levels will build up gradually, with ammonia peaking first. Be patient, and wait until you have had several consecutive days of zero readings before considering adding livestock. When you are certain that all is clear make a 20-25% saltwater change to dilute the resulting levels of nitrate that may form.

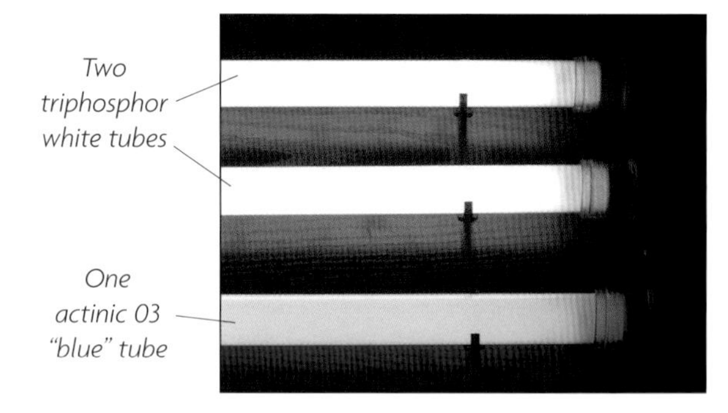

Two triphosphor white tubes

One actinic 03 "blue" tube

18 When using fluorescent tubes it is important they are secured properly into the hood by clips designed for this purpose. Do not use metal clips.

How your aquarium matures

It is vital to understand how the levels of nitrogenous waste products rise and fall during the first few weeks.

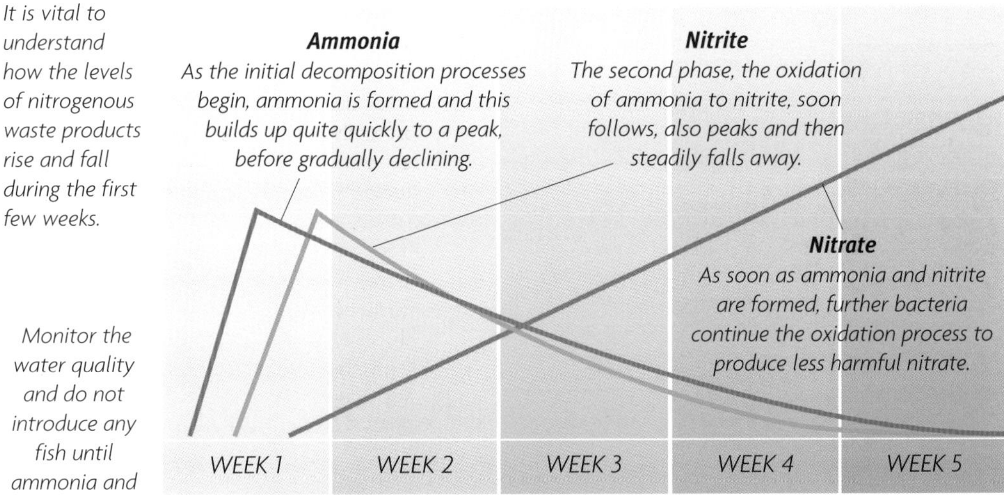

Monitor the water quality and do not introduce any fish until ammonia and nitrite levels have been zero for several days.

19 *Connect the heating and filtration units to the electricity and turn them on. Introducing a "cleaning gang" of hermit crabs and/or snails controls any developing algae growth while the tank is fish-free and lit 12 hours a day.*

Patience at this stage will ensure that your aquarium gets off to a healthy start.

Choosing fish and invertebrates

Several factors must influence your choice of animals for the aquarium; their compatibility, both with their own kind and unrelated species; their diet; their tolerance of captivity; their appearance and even their cost. It is, therefore, very important to research thoroughly all the species you would like to keep.

Obtaining healthy stock

Observe fishes and invertebrates carefully before buying them. They should be alert and free from any sort of damage. Avoid fishes that look thin or that have pinched stomachs, as they often have trouble recovering. Freshwater fishkeepers are used to seeing their healthy fishes swim with erect fins, but many marine fishes naturally keep their fins down, so this is not necessarily a sign of ill-health.

With invertebrates, it is more difficult to tell whether they are in good condition or not. In the case of corals (ideally attached to rock), anemones, and other polyp animals, look out for tissue damage or even decay. The same applies to sponges, but with crustaceans and starfishes it is usually a case of making sure all their limbs are intact. Sea urchins should not be showing any signs of shedding their spines. Generally speaking, it is worth avoiding any invertebrates that appear to be shedding unusual mucus-like substances, unless you know for sure that this is a normal trait in that species.

Stressed fishes will often produce excessive skin mucus that usually has the appearance of white slimy patches. Avoid these fish, even though the temptation may be to transfer them to your own tank, where conditions are much better.

Left: *When faced with the wide range of fishes to choose from at the aquatic store, you should know what to look for, so that you only buy healthy stock. Avoid buying fish simply because you like the look of them; they might not be compatible.*

Many tropical marine fishes and invertebrates are brilliantly colored. Make sure their colors are as they should be and avoid specimens with an unnaturally dull appearance or any with poorly defined markings. Cloudy or swollen eyes, ulcerations, and abnormal swimming or instability are bad signs. In particular, look out for signs of scratching or unusually high respiration, as these indicate parasite infections or other possible ailments, as well as toxic water conditions.

Some species, particularly those caught directly from the wild, may take a while to start feeding in captivity, so it is worth asking if you can observe them eating before you buy. Most aquarium dealers will not have any problem with this. However, one problem arising from this is the risk of the food being regurgitated or the fish defecating in the transit container on the way home and causing an ammonia problem. Once you have seen the animal feed, it is worth leaving a deposit and arranging to collect it the following day. For these reasons it is always worth developing a good relationship and level of trust between yourself and your chosen dealer.

Using the same source for your fishes and invertebrates will help to ensure a consistent quality of livestock and service. If you obtain specimens from a variety of sources it can be difficult to trace where a problem may have originated, and you will not be able to alert the relevant people or expect any one source to take responsibility.

The journey home

Once you have obtained your chosen specimens you will be taking them, presumably, on the final stage of their journey from natural habitat to the aquarium. The best, and standard, way of transporting most fishes and invertebrates is in plastic bags filled with one-third aquarium water and two-thirds air. If your journey home is a long one, explain the situation to the dealer so that you can arrange for oxygen to be injected into the bag instead of air. If the weather is cold, or very hot, you may be able to buy or borrow an insulating polystyrene shipping box to transport your animals.

While in transit, never subject your fish to bright light, but keep them calm in the dark. The plastic bags should be wrapped in paper or kept in a dark paper bag or box. Likewise, do not subject them to sudden movements and vibrations. If traveling in a vehicle, cushion the transit container to prevent excessive vibrations from the engine or uneven road surfaces disturbing the animals.

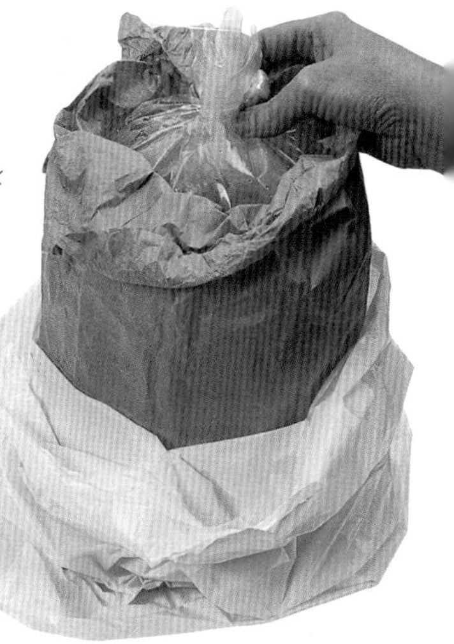

Right: *Once caught from the dealer's tank and bagged up, fish are usually placed into a brown paper bag. Being in dark surroundings usually calms them down and prevents undue stress occurring during the journey home.*

Introducing the fish

Continue to keep the new animals in dimly lit conditions throughout the introduction process. Unpack the bags and float them in your aquarium for approximately 15 minutes to allow any differences in temperature to equalize. If you have had a long journey, there is a risk that the pH of the water in the transit bags will have dropped. Sudden changes in pH are dangerous to marine animals, so gradually mix the transit water with the aquarium water until equalization has been achieved. Open the floating bags and introduce a small amount of aquarium water at a time until the bag is full, making sure it does not sink, then pour two thirds back into the tank and repeat the process. By the time you have filled the bag a second time, you should find that the pH, S.G., and temperature have all equalized. Nevertheless, you must take some tests to be absolutely certain before releasing the animals.

To avoid further stress on the fishes and invertebrates, you should carefully balance the need to acclimatize them with the length of time it takes to do so. Carefully monitoring respiration will indicate to some degree the level of stress a fish is experiencing, but it is much more difficult to assess with invertebrates.

As living rock actually includes living organisms, it is necessary to acclimatize it to your aquarium in exactly the same way you would fishes and invertebrates. The same also applies to algae.

Quarantining new stock

If you are not certain about the health of new animals, it may be advisable to quarantine them in a tank designed for this purpose before introducing them to the aquarium. Always maintain such a tank to the same standard as your main aquarium. Despite the best intentions, it may be argued (with some degree of logic) that any extra transference of animals from one situation to another can cause stress. The extra quarantining stage may be an example of this and will require that you acclimatize the animals in the usual ways when transferring them. Careful handling of healthy stock from a reputable dealer will reduce the number of occasions where quarantining is necessary.

Above: *Equalize the transport water with that in the aquarium by floating the unopened bags in the tank, preferably with the lights off, as this will help the fishes or invertebrates stay calm during this process.*

Left: *After the bags have floated for 15 minutes you can open them and start gradually mixing in some aquarium water as described in the main text. Doing this will equalize differences in pH and salinity and "fine tune" the temperature before release.*

The first days

After the animals have been successfully introduced into the aquarium, leave them with the lights off, but not in total darkness, for about an hour. After this the lights can be turned on in stages, and it is often a good idea to offer a very tiny amount of food. For many animals, eating when in a new situation is what is known as a displacement activity and the fact that food is available helps them to settle. Obviously, the presence of healthy living rock will provide a number of suitable food items in the form of microorganisms and algae. However, some fishes may not feed for a couple of days while they are settling in and, in most cases, this will not cause undue harm. The most notable exceptions are surgeonfishes and seahorses, which must keep eating virtually constantly.

Whether you are introducing animals to a newly set up aquarium or an established system, add only one fish at a time to avoid overloading the capabilities of the biological filtration. Gradually the bacteria population will catch up with the extra loading and you will be able to make another addition. For corals and other invertebrates, you can count a cluster of polyp animals on a rock as the equivalent of one animal, providing it is not a massive piece. In fact, invertebrates can be added in pairs at weekly intervals.

It is worth bearing in mind that any animals known to exhibit particularly aggressive territorial behavior are best introduced last of all. If established in the tank any earlier, these animals could mercilessly attack newcomers that are introduced after them.

Introducing invertebrates

The same general guidelines that apply to fishes should be observed when introducing invertebrates to the aquarium. These include sea anemones, crabs, shrimps and prawns, tubeworms, featherdusters, starfishes, sea cucumbers, sea slugs, and sea urchins. To keep invertebrates you will need to learn different skills; many are normally sedentary, hardly moving from their coral or rockbound sites. Feeding can be a collective affair for filter-feeding species, but you will need to feed sea anemones by hand, placing morsels of food into their tentacles with tweezers.

Invertebrates will reproduce in the aquarium; sea anemones, for example, will simply divide, or expel young forms to increase their numbers. Hermit crabs are fascinating to observe when they hunt for larger "premises" as they outgrow their former dwellings. Even at night there is activity in the aquarium; many of the soft corals and similar animals emit small points of light in the darkness.

Above: *Avoid placing individual corals and anemones too close to each other and never touch their living tissue with your fingers as you position them, as this may cause serious damage to the animals.*

Left: *The same approach applies to living rock, algae and corals. Take care not to damage other invertebrates when placing new items into your habitat formation. Place them onto the rockscape in stable positions to prevent other tank inhabitants from easily dislodging them.*

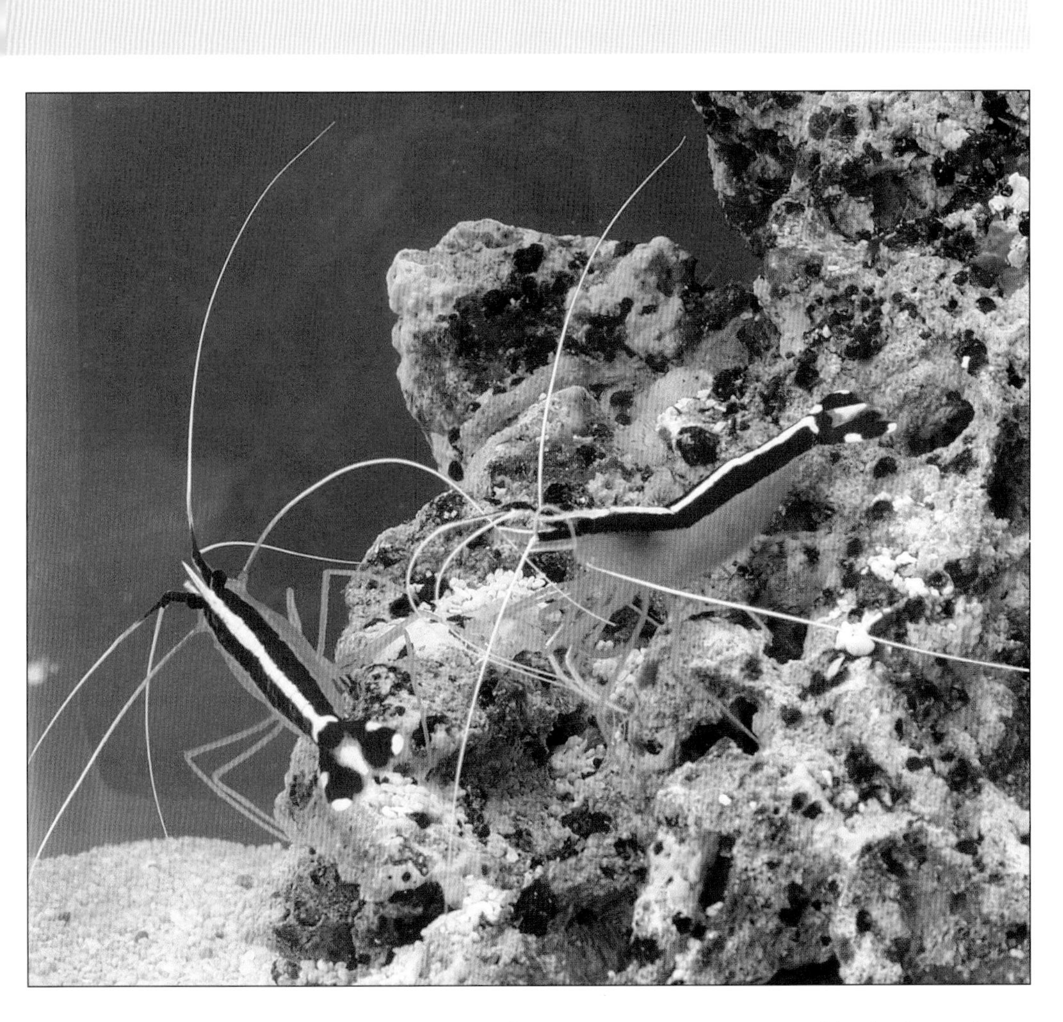

Above: *A pair of cleaner shrimps. All kinds of invertebrates require the same type of careful treatment as marine fishes when being introduced to the aquarium.*

The finished display

The finished aquarium, including the initial livestock. Only introduce one fish or invertebrate at a time, unless they are small enough to allow more – please use good judgment! Always monitor the water quality before making any further introductions to avoid overburdening the filter system and in turn stressing the animals.

In addition to playing an important role in water-quality control, the living rock habitat supports its own community of living organisms but also provides retreats for other aquarium subjects and a base for adding live corals, anemones and other sessile creatures.

Yellow sailfin tang, a hardy fish ideally suited to this type of aquarium setup.

Compatibility

It is important to consider the compatibility of both the fishes and invertebrates you intend to keep together in the same aquarium. Nature seems cruel when you think about a large predatory fish eating a smaller species, which in turn may eat a yet smaller fish or invertebrate; and certain invertebrates are quite capable of devouring fishes. Sadly, tank inmates do not necessarily become docile in the home aquarium where food is more easily available, so food chain trends continue.

If you are not certain about the normal lifestyle of a species, work on the principle that if one fish is bigger than the others in the aquarium and has a large mouth, then the chances are that smaller fishes will be eaten. Groupers, lionfishes, anglerfishes, and moray eels are typical of those in this category. Likewise, invertebrates with large claws are potentially going to do some serious damage, and anemones with powerful stinging cells can easily catch slow, weak-swimming fishes, such as seahorses.

Less obvious threats are butterflyfishes that eat the stinging cells of corals, filefishes that often bite out the eyes of less robust species, and triggerfishes that have jaws adapted to eating spiny sea urchins. Another consideration is that when a fish that naturally lives in a shoal is kept as an individual, it may change its defense tactic by becoming aggressive and territorial, because it can no longer rely on the safety of numbers. The best strategy against predation in the aquarium is to read about the lifestyles of the animals you are interested in before obtaining them.

To avoid other types of aggression, it is a good idea to learn how to tell sexes apart. In the case of damselfishes, for example, males that have established a nest site will fiercely fight with one another, and other fishes that get too close for that matter, while females are happier living together out in the water column. Male damsels often have color in their fins, but the females' fins are transparent.

Corals, anemones, and marine algae

Living hard corals (or stony corals), the reef-building corals, are controlled by C.I.T.E.S. (Convention for the International Trade in Endangered Species) but licenses to import them can be obtained. Clams (*Tridacna* spp.) are also controlled, but soft corals, anemones, and other polyp animals are not. Commercial coral propagation farms now exist, and you should try to buy pieces that have originated from such sources whenever possible. Many aquarists have successfully propagated their own corals and swapping pieces with fellow enthusiasts is quite common. You can also obtain artificial coral "skeletons" molded in resin (see page 45).

Adding corals and related animals to the marine aquarium is not as difficult to achieve as it was some years ago. However, hard corals are still not for the inexperienced and you should begin with other, tougher, related polyp animals. Always bear in mind that certain species of fishes will eat corals and other invertebrates, even those found on living rock, so some forethought will be required when considering what mix of animals you wish to keep.

Marine algae

The sheer mass of algae on a reef is quite enormous. Apart from the more recognizable types of macro-algae that grow on one substrate or another, there are the zooxanthellae that live within the tissues of corals, particularly hard corals, and clams. It is because of these algae that corals can only survive in areas of very strong sunlight. Like other plants, zooxanthellae need light to photosynthesize.

By including macro-algae in the reef aquarium, you can provide the plant element that many aquarists miss if they have moved from keeping a freshwater aquarium to a marine tank. Macro-algae can also be very useful for marine fishes that like to graze, such as herbivorous surgeonfishes. The algae derive much of their nourishment from waste products in the aquarium water, but you may have to add trace elements, fertilizers and even extra carbon dioxide, which is perhaps the most important nutrient for plants. One drawback of achieving a lush growth is that it may suddenly "crash," releasing toxins into the aquarium water as it dies back. To avoid this, harvest your algae periodically. Macro-algae, especially some of those in the genus *Caulerpa*, are capable of rapidly taking over the tank, as their daily growth rate is quite phenomenal when conditions are right.

Above: *Caulerpa comes with many "leaf" shapes, but these grapelike clusters instantly identify the species shown here as* Caulerpa racemosa. *This is an attractive and relatively slow-growing form.*

Left: *When the normally extended polyps of this leather coral are retracted, all that is visible are these leathery-textured fingers. There are many forms available for the hobby, and they thrive in brightly lit tanks.*

Above: *Similar in appearance to mushroom coral, the polyps of finger coral* Sinularia *are displayed on separated lobes rather than on an undulating plate.*

Left*: Large anemones, such as this giant carpet anemone,* Stichodactyla gigantea, *require intense lighting if they are to reach these proportions in a marine aquarium.*

A reef aquarium

The ultimate type of aquarium for most marine hobbyists is a reef aquarium populated by an abundance of corals and macro-algae. Such aquariums often contain very few fish, as the main focus is on invertebrates. A reef aquarium is often best decorated by using a base of tufa with live rock placed on top. Alternatively, glass shelves can be siliconed into the aquarium, with live rock placed on top. This method has the added advantage that valuable water space is not displaced by tufa or ocean rock.

When buying live rock, always ensure that it has been properly treated before adding it to an established aquarium. Uncured rock will give off high levels of toxic ammonia. Once it has become established, the rock will develop attractive algae, coral, and sponge growths. Reef aquariums are best illuminated by metal-halide lamps, in order to provide the correct intensity and light wavelengths for the zooxanthellae required by many corals and other invertebrates to grow (see page 39).

Water quality in a reef aquarium must be of a high quality, and it is fortunate that once it has become established, living rock is capable of nitrate reduction in the aquarium. Not only do hard corals require the correct lighting and good water quality, they also benefit considerably from a high water flow, provided by suitable powerheads. Furthermore,

***Right**: A successful marine aquarium will allow each of the inhabitants to establish its own territory within it. This means providing plenty of safe retreats.*

***Left:** In the modern reef aquarium, the emphasis is more often centered around coral growth than on the fish population.*

powerheads provide a flow of water that washes away any detritus from the surface of corals, and prevents them from rotting away.

Given the amount of equipment generally required to maintain a reef aquarium, it is a good idea to use a sump system and to house the protein skimmer, heating units, filtration, and recirculation pumps outside the aquarium. The only pieces of equipment required inside the aquarium are the powerheads to provide strong water currents, which can easily be hidden among rocks.

Reef tanks with a capacity of less than 100 gallons (450 liters) are not recommended, especially for less experienced hobbyists, as the exacting water requirements of hard corals are extremely difficult to maintain in a small aquarium.

Feeding marine fish

Ideally, the best diet for any animals in captivity should be based on the food they would have enjoyed in nature, so we should not be surprised that the best diet for the majority of marine fishes is fish! The exceptions to this general rule are fish that are herbivorous by nature or those whose tiny mouths mean that they feed on microscopic animals such as plankton.

The shape of a fish's head or snout is a good clue as to how it obtains its food. For example, butterflyfishes often have elongated snouts, ideally suited to picking food from crevices within the coral reef. Parrotfishes have strong fused teeth with which they crunch up the coral in search of food. The fearsome jaws of groupers and lionfish are obviously designed for copious food gathering. The mouths of blennies and gobies are located at the very front of the head, and the flat ventral surface of the body brings the mouth into efficient proximity with the food-laden substrate.

A feeding regime

Other considerations are when – and how often – to feed fish, and here again, observing nature will help the fishkeeper. Large-mouthed, predatory fish may gorge themselves initially and then fast for a

Frozen foods

Whole cockle
This natural food is accepted by all marine fishes.

Marine mix
This is a mixture of various natural marine invertebrate and fish meats.

Krill
A nutritious food for larger fish; break up for small fishes.

Fish
These small fish make excellent "one-gulp" food for larger fishes.

Shrimp
Many wild-caught foods are irradiated to destroy disease pathogens.

Above: *Frozen foods are sold in single slabs (simply break off pieces as required) or in individual push-out "servings." Thaw frozen foods before use to prevent fish eating ice.*

Ease of feeding in the aquarium

Group	*Easy*	*Reasonable once feeding*	*Difficult*	*Difficult, need live foods*
Angels		●	●	
Basses	●			
Blennies	●	●		
Butterflyfishes	●	●	●	●
Cardinals		●		
Catfishes	●			
Clownfishes	●	●		
Damsels	●			
Eels	●			
Filefishes			●	
Gobies	●			
Jawfishes		●		
Lionfishes		●		●
Seahorses			●	●
Squirrelfishes		●		●
Surgeons	●			
Tangs	●			
Triggers		●		
Wrasses	●			

Feeding for groups with multiple entries depends on species.

few hours; herbivorous grazers appear to be active as long as there is daylight. However, the majority of fishes in the aquarium will be satisfied with one or two regular feeds each day. Do not neglect the needs of the more nocturnal species. They should be fed after the aquarium lights have been switched off for a little while.

On balance, the fishkeeper would be well-advised to underfeed, rather than to give food every time the fish appear to indicate that they require it by their behavior at the front glass. You will know if you are feeding the correct amount of food because the fish will consume it all within a few minutes and there will be none left to pollute the water.

Feeding marine animals other than fish in the aquarium may require special techniques. For example, in the case of filter-feeding invertebrates in a reef tank, it may be best to turn off the filtration system briefly, so that the action of the filter does not remove the food from the aquarium before the invertebrates can make use of it.

Feeding marine fishes

All suitable foods for marine aquarium fishes should be processed in one form or another. Unless the source is meticulously screened, providing live food runs the risk of introducing undesirable pathogens to the aquarium. There is a school of thought that believes that offering live fish, such as goldfish, to the larger, more predatory marine fish in the aquarium may lead to the marines "learning" to be predatory toward their smaller tankmates.

Given that the ideal diet should be marine based, you will find all manner of frozen fish- and shellfish-based foods available from your aquatic retailer. These have probably been processed by using gamma irradiation to safeguard against the introduction of disease. The freeze-drying process may also be used to preserve aquatic animals as prospective food, and such products are also readily available.

Much research has been done by the fish food manufacturers, and a whole range of recipes suitable for marine fish will be available in many forms: flake food, tablets, granules, or even sticks. While such foods may appear expensive, it is false economy to buy a large amount, especially if you only have one aquarium. Once the food container has been opened, the quality of the food (and the important vitamin content) will gradually deteriorate, thus denying your fish all the benefits the manufacturers have engineered into the food.

Feeding herbivorous marine fish

Dealing with the needs of herbivorous species is not as difficult as you might imagine. If the lighting level is sufficiently strong, a healthy growth of algae should soon manifest itself over the surface of the aquarium walls and decorations, and herbivorous species will be only too glad to graze on this green matter. Alternatively, culture algae-covered rocks in a separate tank standing in a sunny position and filled with discarded aquarium water following a water change. Transfer each "green" rock to the main aquarium in rotation, and replace the "grazed upon" rocks in the culture tank. This should ensure a steady vegetable diet for those fish that need it.

Provide further vegetable matter in the form of peas, spinach, and blanched or lightly bruised lettuce leaves. Clamp the leaves between the two halves of a magnetic algae scraper to prevent them circulating around the aquarium in the water currents.

Right: *This decorated firefish* (Nemateleotris decora) *is a midwater feeder and clearly has no problems in taking food as it falls through the water. Here, it is about to "capture" a fragment of thawed-out shrimp.*

Flakes and granules

Flake foods fall slowly through the water and are ideal for most midwater and upper water level species.

Mixed flakes

Brine shrimp

Granular food
This sinks quickly for the benefit of bottom-dwellers.

Tablet food
Excellent for sticking on the tank glass.

Above: *To prevent green foods becoming stuck in filter inlets, secure them to the aquarium glass using a sucker and clip.*

Freeze-dried foods

Based on fresh natural foods, most freeze-dried portions can be stuck onto the interior glass panels at any level.

Freeze-dried krill

Freeze-dried river shrimp

Freeze-dried brine shrimp

Routine maintenance

Once your aquarium is completely up and running, the time taken by routine maintenance is relatively small, yet it is something you can still enjoy.

Routine checks

There are several checks that you should make on a regular basis. First and foremost are daily temperature checks, along with a "head count" of your aquarium subjects. Counting fishes is best done at feeding time, when they tend to be the most visible, although some invertebrates may be more difficult. Over time, you will develop an "eye" for what is right and wrong, so that you can tell almost instinctively if any animals are more than just missing. Do your best to locate them and immediately remove any that have died. If an animal dies and the cause is unlikely to have been old age, you must try to discover the cause and take action accordingly. Marine fishes and invertebrates are generally sensitive to disturbances and the potential stress they cause, so do your best to use your eyes rather than searching around the tank with your hands. Some species are capable of jumping, so if an animal is not evident in the aquarium, it is worth looking on the floor, especially if the tank is not covered.

As the aquarium gradually becomes more established, its subjects will settle down into their natural behavioral patterns. If you take time to observe the animals' normal activities, you will soon be able to recognize any irregularities that may be early indications of trouble.

In a fully established aquarium, you should check pH weekly, making sure it remains stable at around 8.3 – but no higher – and no lower than 8.1. If it

Testing for ammonia, nitrates, and nitrite

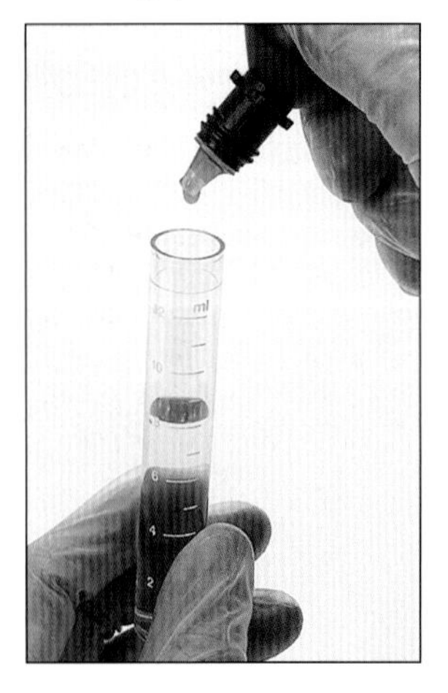

Testing the water involves adding chemicals to a measured sample and comparing the color change to a printed chart. Some tests involve adding two or three chemicals in stages. Allow the correct time period to elapse between adding reagents. Wear protective gloves when using chemicals, as they may cause skin irritation.

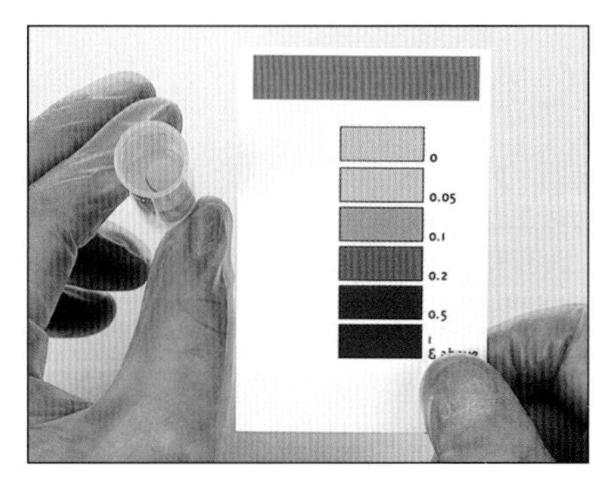

Above: *Any indication of the presence of nitrite does at least mean that the bacteria are doing their job in the conversion process but, like ammonia, nitrite is still very toxic, and a zero reading is what you need.*

starts to drop, adjust it by adding buffering materials (see pages 33). The situation may require some water renewal, especially if the pH continues to drop after you have attempted to adjust it. Siphoning off debris from the substrate on a weekly basis will go a long way in preventing a decline in water conditions, especially pH. Most of this debris is likely to be organic, so as it breaks down it will be releasing carbon dioxide into the water and affecting its buffering capabilities.

Take a specific gravity reading once a week and test for ammonia and nitrite levels once every two weeks. It is also a good idea to keep a check on the nitrate and phosphate levels weekly. Fishes generally appear to have a higher tolerance of nitrate than many invertebrates, and experienced marine aquarists have suggested 20 mg/liter is an acceptable upper limit. However, it is always best not to make any compromises and try to keep nitrate levels below 5 mg/liter.

It is important to carry out regular partial water changes. Do not change too much at any one time as it may cause a sudden drift in environmental conditions and cause some stress to the animals. Around 10% change each week is good, although smaller amounts on a daily basis would be more efficient, but far less convenient, for most aquarists. As already discussed elsewhere, never mix new supplies of saltwater directly in the tank once there are living organisms in it. Occasionally, it may be necessary to replenish due to evaporation loss. As only water evaporates, leaving the salts behind, you must only replenish by using freshwater. Always use the best quality freshwater available to you for this purpose.

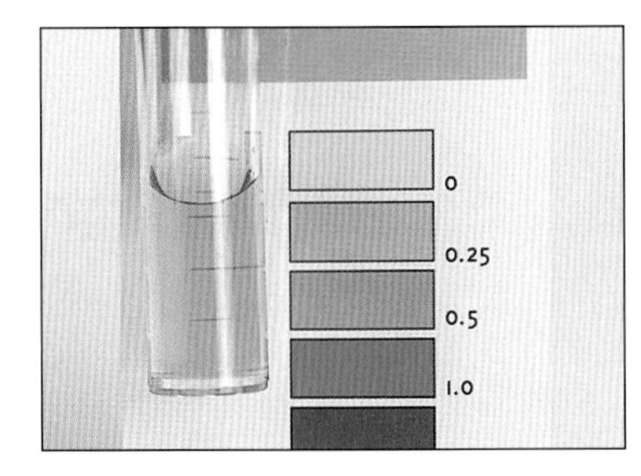

***Above:** When testing for ammonia, the aim is to achieve a reading of zero. Not until this reading is consistently at the lowest possible value, is it safe to introduce any fish into the aquarium. Test the water regularly.*

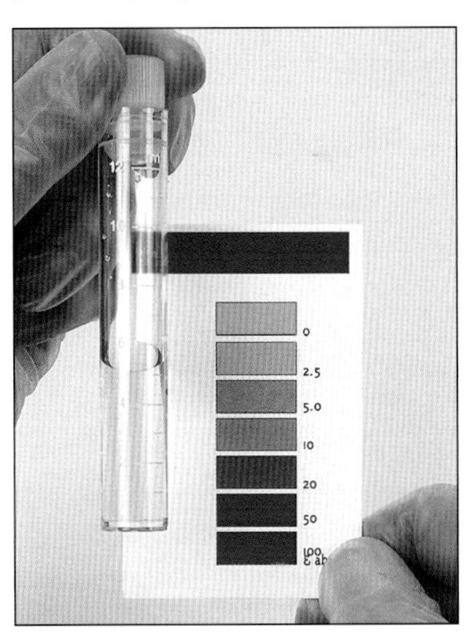

***Left:** With some original nitrate already present in the mixing water, readings are likely to be high. Although nitrate can be tolerated by some species, it should not be allowed to exceed 10 mg/liter. Here, the reading is very low.*

Routine maintenance

Occasionally things can go wrong. If a heater fails, causing a serious drop in temperature, it is a good idea to float plastic bottles filled with hot water in the aquarium, taking care that the displacement of aquarium water does not overflow. If you own an enamel pan, you could also heat up some aquarium water and return it to the tank very carefully and gradually. In the meantime, either repair the heater or replace it with a new unit – the latter being the most likely course of action.

On the other hand, should your heater's thermostat stick in the "on" position, causing the temperature to rise, switch off power to the unit and apply extra aeration. In a reverse of the method described above, you could fill plastic bottles with ice in order to reduce the temperature gradually.

Phosphate test

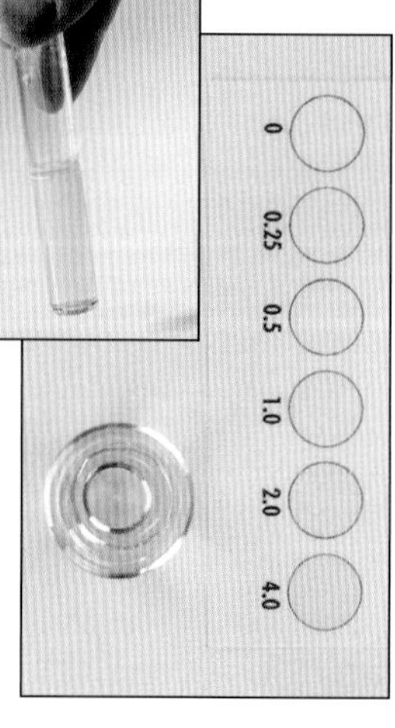

Add phosphate reagent to a sample of tank water and mix. After five minutes look down the tube to compare the color of the solution with the chart.

Should your electricity supply fail, such as in a power cut, the relatively large body of water in the aquarium can act as a heat reserve and will take a while to cool down. Lagging the tank with insulating material will help if there is concern that the power outage will last a long time. If you have other means of heating water, you could use bottles as described above. It is more critical to keep the filters running and to provide extra aeration. If the nitrifying bacteria die off due to a lack of oxygenated water, you run the serious risk of deadly ammonia buildup. In such an event, one or two battery-powered airpumps would be very valuable investments and could maintain both water movement and the filter system until power is restored.

Backup systems are worth considering. To avoid disaster from a sticking thermostat, you could connect a second one in line to the power supply. Set the second thermostat at a couple of degrees higher than the first, so that if the main one sticks, the other one will turn off the heater. Otherwise, you could install two units as a precaution against one failing and causing a temperature drop. A further refinement would be to install a battery-operated buzzer alarm to indicate power failures.

Look on the bright side

Finally, remember not to look on routine tasks as chores to be avoided; most of them should be part of the pleasurable task of providing your aquarium subjects with the care and attention that they undoubtedly deserve. They will repay you in the best manner possible – with a living picture of beauty and color that is both entertaining and educational. As a hobbyist, this is your reward.

Practical tips for a healthy aquarium

Choose healthy stock and quarantine all new additions. Avoid stressing the fish in any way during transportation and subsequent handling.

Bear in mind the feeding habits of different species, their compatibility with fish of the same or different species and their eventual mature size.

Remove uneaten food at the earliest opportunity. Acclimatize fish to any new food over time.

Learn to recognize symptoms of impending water problems: frothy, cloudy, yellowing, and smelly water are all signs of deteriorating conditions.

Keep cover glasses clean. Salt spray soon renders them opaque, which prevents the full light intensity reaching right down into the water.

Make all changes to the water conditions gradually. Powerheads or other motorized equipment may be left "high and dry" as water levels fall; turn them off during water changes but turn them on again!

Do not neglect regular partial water changes. Changing 20-25% each month is a rough guide; an invertebrate tank may require more frequent changes.

Check that the waterflow rate from filters remains high. Clean external mechanical filters and replace (or rinse out and reuse) the filter medium together with the activated carbon frequently – at least once a week if necessary. Wash filter media in some aquarium water to avoid killing any bacterial colonies. Fit isolating taps in filter hoses to lessen the risk of spillage when cleaning filters.

After cleaning external power filters, make sure that the hoses are securely attached – a filter pump will just as easily empty a tank as filter it!

Remove algae from the front glass with a nonmetallic scraper; nylon or plastic scourers are very effective for this purpose. Algae scrapings can be fed to herbivorous fish in less algae-covered tanks. Thin out excessive macro-algae growths, as a sudden "algae death" could cause pollution.

When treating fishes with disease remedies, follow the maker's recommendations. Most treatments are designed to be added to the whole aquarium, but remember that copper-based cures will kill most invertebrates in the tank. (This is why it is difficult to keep fishes and invertebrates together. A practical ratio would be 80% invertebrates to 20% or even fewer, fishes.) Do not mix medications. Sterilize all equipment after use, and do not share a net between two tanks.

Replenish evaporation losses in the aquarium with freshwater.

Health care

Even assuming that your fish were in the best condition when they became your responsibility, it is nevertheless a fact of life that, at some time or another, you will have to face up to an outbreak of disease in the aquarium.

Look and learn

Get into the habit of learning from what you see. Once you can recognize each fish's normal behavior, you will soon be able to detect any abnormalities, such as erratic swimming, sulking, and hiding away. Concave profiles (especially on a new fish) should be regarded with suspicion.

Remember that many marine fish, such as the birdmouth wrasse, swim with their fins folded down as a matter of course, and in any case, some species lack some fins found on other fishes!

First things first

When dealing with disease, you should not only identify the ailment but also look for the causes. Always check the aquarium conditions (especially water parameters) first. Attention to aquarium hygiene, modest stocking levels, and quarantining all new stock will do much to ensure that your aquarium remains stress- and disease-free.

Copper remedies

It is not possible to treat fishes in the aquarium if invertebrates are also present. Although fish respond well to copper-based remedies, invertebrates have no tolerance to them at all and will be killed.

The treatment tank

One reason for using a separate treatment tank is that some medications can affect the nitrifying bacteria, with the result that the main tank needs time to "remature" once treatment is complete. A further problem occurs if you are using filters that use carbon, which removes the treatment from the aquarium water. Always remove carbon from the filtration system before dosing with treatments.

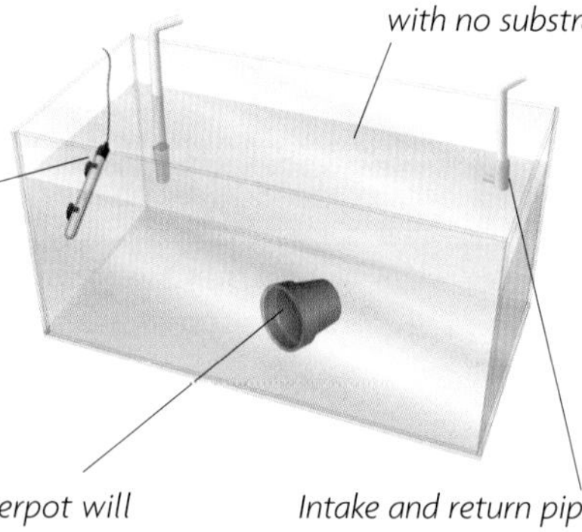

Disposing of fish humanely

At some point in your fishkeeping career, you may find it necessary to kill a fish because of illness or old age. If a sick fish is beyond saving, the best thing is to dispose of it humanely. If you can bring yourself to do it, the quickest method is to sever the spinal cord behind the head with a sharp knife. Alternatively, obtain the fish anaesthetic MS222 from a veterinary surgeon or pharmacist and leave the patient in a solution of this for several hours. Never flush a fish down the lavatory, throw it on the floor, or place it alive in the freezer. Freezing affects the capillary blood vessels just under the skin, causing the fish great pain before it loses consciousness.

Right: *Accurate dosing is essential, so it helps to know the exact volume of water in your aquarium. Pre-mix medication in a little tank water before adding it to the treatment tank. Never add another remedy immediately should the first one not work, as combining several remedies may be toxic.*

Many common problems have easily detectable, external, visible symptoms, such as tiny white spots, etc. These can be treated successfully by using commercially available treatments. Internal disorders, with external symptoms that may not become apparent until the disease has progressed too far for successful treatment, are more difficult to deal with and usually result in the loss of the fish concerned. Some remedies are not always available worldwide due to differing regulations on drug sales.

Tread carefully

Treat your fish in a methodical way and make notes if necessary. Be confident of your diagnosis before embarking on a course of treatment. In an emergency, some parasitic attacks can be dealt with by a simple freshwater bath. Although it sounds contradictory, a bath in freshwater for some minutes (less if the fish shows signs of discomfort) can bring relief to marine fish.

It might not be possible to arrive at an accurate diagnosis immediately, since there may be more than one explanation for any given symptom. For example, fishes with breathing difficulty may not be the subject of a parasitic attack at all; it may be that there is simply not sufficient oxygen in the water available to them. (That is why it is so important to check the water conditions first.) However, if the fish's gills really are covered with parasites, then this will cut down the efficiency with which they can extract oxygen from the water and results in their rapid, gasping breathing actions.

Health care

Detecting and treating health problems

Symptoms	***Cause***	***Treatment***
Opaque patches on skin.	Chilodonella.	*Antiparasite treatment effective as a host is needed; an uninhabited aquarium should be clear after 5 days.*
Triangular spots.	Benedina.	*Proprietary treatments effective. Use brief freshwater baths with caution.*
White spots, fish scratch against rocks.	Cryptocaryon irritans.	*Use copper-based remedy in separate tank if invertebrates present.*
Dusty, velvet-like spots; fish scratch, gills inflamed, breathing affected.	Amyloodinium ocellatum.	*Proprietary remedies effective but avoid copper-based treatments if invertebrates present.*
Emaciation – but eating regularly.	Ichthyosporidium *or* tuberculosis.	*No effective cure. Antibiotics may help; isolate sick fish.*
Eyes – Bulging, pop-eye.	Exophthalmia.	*No definite remedy but fish often not distressed by condition.*
Fins – Ragged with wasted tissue.	*"Finrot," bacterial infection, usually secondary infection.*	*Use copper-based remedy. in separate tank if invertebrates present. Improve water conditions.*
Gills – Open, inflamed.	*Gill flukes,* Dactylogyrus.	*Brief formalin bath.*

Symptoms	***Cause***	***Treatment***
Cotton wool-like tufts.	*Saprolegnia fungus.*	*Immediate partial water change; brief freshwater baths and antibiotic treatment. Improve water conditions.*
"Cauliflower" growths.	*Lymphocystis.*	*Fish either waste away or recover completely.*
Slimy feces.	*Diarrhea; poor diet.*	*Raise tank temperature a few degrees; starve fish for two days, then feed high roughage foods.*
Swellings.	*Bacterial infection or, in severe cases, edema.*	*Use copper-based remedy in separate tank if invertebrates present.*
Wounds.	*Physical damage. Poor handling. Bullying. Ulcers caused by* Vibrio *bacteria.*	*General antibacterial treatment, local swab medication. Consult veterinarian.*
Rapid breathing coupled with dashing about, dull eyes.	*Poisoning.*	*Immediate partial water change with total change later if needed.*
Gaping gills.	*Gill flukes (see "Gills").*	*Brief formalin bath.*
Listlessness.	*Associated with several other disorders.*	*Diagnose and treat accordingly.*
Loss of balance, cannot maintain position in the water.	*Swimbladder disorder.*	*Isolate fish in warmer water. Try medicated foods.*

Breeding marine fish

The art of breeding marine fishes in captivity is still in its infancy. However, well over 100 species have been bred in aquarium conditions and some of these are available commercially. Successfully breeding and rearing a species must be the ultimate achievement for any marine aquarist, and the results are crucial for the future of the hobby and for conservation. The knowledge gained from such experience will in time help to reduce the demand for wild-caught specimens and the conservation of natural habitats. Just as importantly, this information could be used to teach those who are currently collecting for the aquarium hobby to set up breeding stations in the fishes' countries of origin as an alternative means of income should the demand for wild stock diminish. There are tried and tested methods for breeding and rearing some marine species, but there is still a great deal of experimentation to be done for most.

Anemonefishes – egglayers

Anemonefishes are among the most popular of marine aquarium fishes and, although not the hardiest, they are probably the most likely to breed in a home aquarium. The common clown anemonefish, Amphiprion ocellaris, *has been bred in captivity over and over again since the early 1950s and has fascinated even novice marine fishkeepers with its characteristic breeding behavior.*

Right: *A pair of common clownfish* (Amphiprion ocellaris) *guard the eggs they have laid on a tile in their otherwise bare tank. In the wild, they clean a flat rock surface on which to lay their eggs.*

A clownfish breeding tank

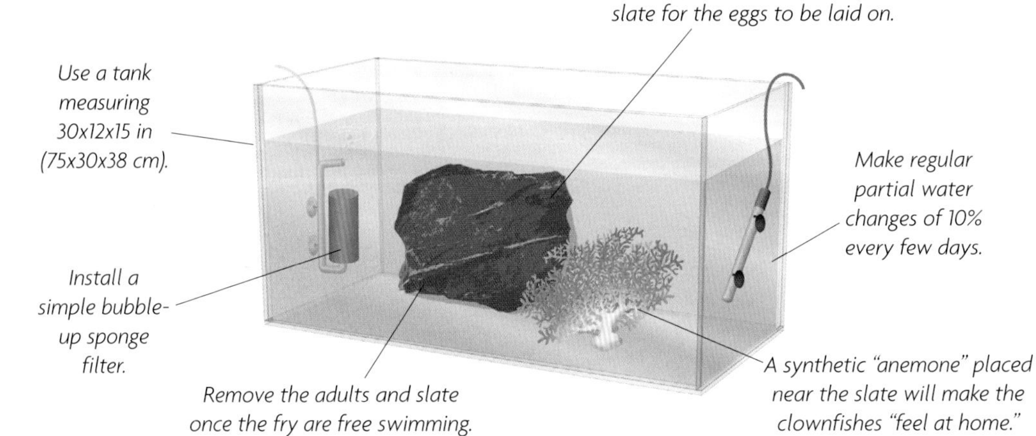

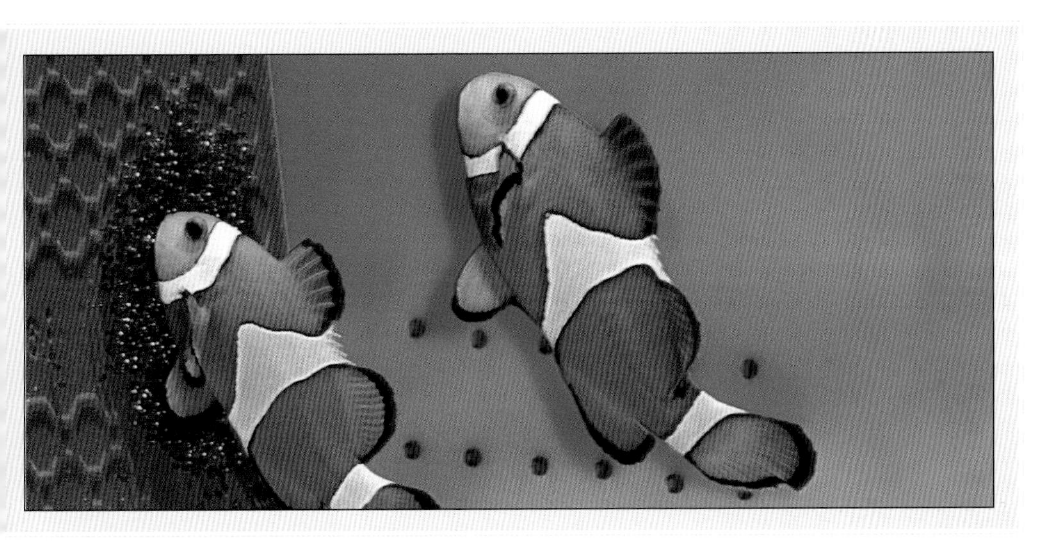

Cardinalfishes – mouthbrooders

One cardinalfish has captured a great deal of interest in recent times – the Emperor, or Banggai, cardinalfish, Pterapogon kauderni. *This species appears to be very restricted in its distribution, being found at only a couple of locations around the Banggai Islands off the coast of Sulawesi in Indonesia. Because of its limited distribution, there has been some concern that collecting for the aquarium hobby will impact on the wild population, so breeding the fish in captivity is a worthy initiative. Fortunately, they reproduce readily in the aquarium and it is hoped that in the near future all specimens offered for sale will be captive bred.*

Above: *The male mouthbroods the fertilized eggs. Once hatched, the free-swimming fry take newly hatched brine shrimp as a first food.*

Part Two

Introduction to fish profiles

Although nearly three-quarters of the earth's surface is covered by salt water, relatively few marine fishes are kept in the aquarium. And of these, by far the most are tropical species. Nevertheless, such is the appeal of this small group of fishes that the marine fishkeeping hobby has flourished and continues to attract more hobbyists around the world.

The most striking tropical marine fishes are native to the coral reefs and coastal waters, where collection is quite easy. Fishes from the deepest waters usually grow too large for the aquarium, and also present too many collection and transportation problems. The majority of suitable fishes come from the Indo-Pacific Oceans, the Caribbean area of the Atlantic Ocean, and the Red Sea.

The water conditions on a tropical reef are extremely stable – the water is well oxygenated and almost completely free from waste products due to the constant cleansing action of the sea. The water is relatively shallow and this means that the fishes are quite used to fairly high light levels. One distinct advantage of simulating these brightly lit conditions in the aquarium is that growth of green algae (usually shunned by freshwater fishkeepers) can be encouraged, much to the appreciation of the fishes, many of which are herbivorous by nature.

Coral reef fishes are extremely territorial, each fish's chosen "living space" in nature being much larger than that possible in the average indoor aquarium. This makes keeping shoals difficult in many cases because the fishes are intolerant of other members of the same species, although they rarely regard different "neighbors" as a threat.

These are the challenges facing the tropical marine fishkeeper: maintaining clean, stable conditions; providing enough space; and choosing compatible species. Achieve these successfully and you can then sit back and admire these wonderful "living jewels of the sea."

Anemonefishes & damselfishe

The family Pomacentridae contains two groups of fishes kept by most marine fishkeepers as initial stock fish when first setting up the marine aquarium. The reasons for this are easy to understand; the fish are hardy, inexpensive, and readily available. Because of their hardiness, they can tolerate early introduction into a newly setup aquarium that may not be sufficiently established to support more delicate species. However, this is no excuse for the impatient fishkeeper to skimp on preparing the aquarium properly before introducing any fish.

The family consists of two distinct groups of fishes, both of which need to feel secure within the aquarium. Anemonefishes, as their name implies, seek refuge within the tentacles of a sea anemone, while damselfishes tend to congregate around the branches of any stands of coral. Each group has an individual swimming style: anemonefishes waddle about, while damselfishes tend to have a more vertical bobbing action.

Caring for these fish is straightforward. They will accept most foods and are not too antagonistic toward other fishes in the aquarium. Damselfishes may squabble among themselves, so provide an aquarium large enough to give each fish a reasonable amount of territory. Several species are likely to breed within the confines of the aquarium, depositing and guarding their eggs in much the same way as freshwater cichlids.

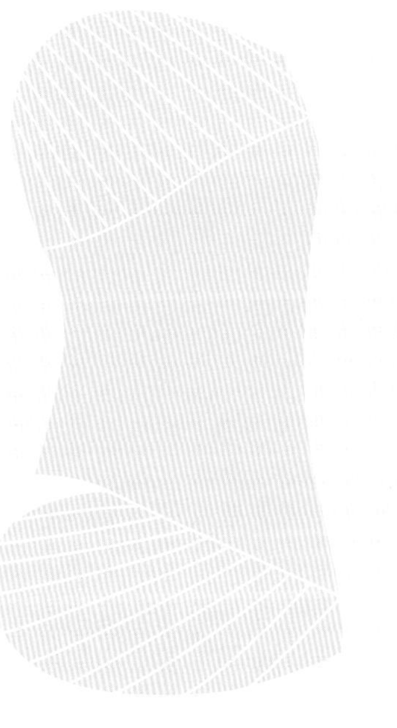

***Above:** This spinecheek anemonefish* (Premnas biaculeatus) *is the largest of the anemonefishes, but still seeks the reassurance of safety among the tentacles of a sea anemone.*

COMMON CLOWNFISH • *Amphiprion percula*

For most people, the definitive anemonefish is the common clown. It is able to live within the tentacles of the sea anemone with impunity, because the mucus on its skin prevents the sea anemone's stinging cells from operating. The ability to dash into the tentacles when pursued by predators stands the clownfish in good stead, as its pursuers will not be similarly immune. Some authors are of the opinion that as a reward for acting as a safe haven, the sea anemone benefits from particles of food accidentally (or perhaps purposely?) dropped by the returning clownfish. Although this mutual friendship may be classed by some as symbiosis, neither of the animals is strictly dependent on the other, and perhaps the more correct term for this association would be commensalism.

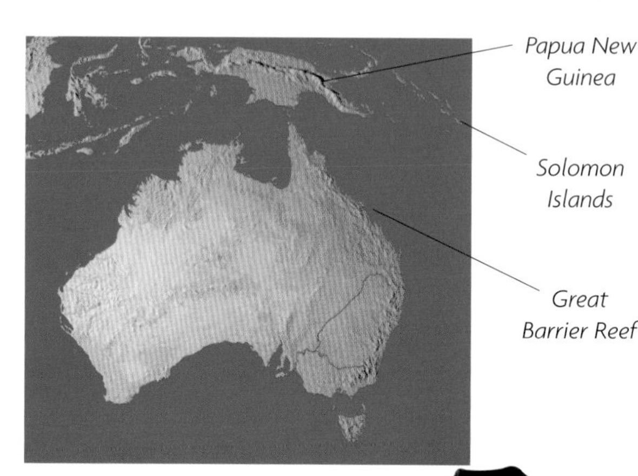

Origins

One of the most widespread species in the genus, ranging from Papua New Guinea to the Solomon Islands and Australia's Great Barrier Reef.

Breeding

Eggs are deposited and fertilized on a firm surface such as a nearby rock. The parents guard them until the fry hatch. Clownfishes are often increasingly available from captive-bred stocks, thus lessening the need for wild-caught specimens.

The distinctive body pattern is shared by the almost identical species, A. ocellaris, the false clown anemonefish. However, it is easy to distinguish between them, because the common clownfish has the wider black markings.

Size: *4.3 in (11 cm); females are generally larger*

Fishkeeping data

Number per aquarium: Two to three, plus a suitable sea anemone.
Community/species tank: Community.
Swimming area: Middle and lower levels.
Food: All prepared foods, including freeze-dried and frozen, plus some vegetable matter.
Compatibility: Peaceful.
Availability: Regularly available (tank-bred and wild-caught).
Captive breeding: Possible.

Clownfishes and anemones

Buying clownfish complete with an associated anemone is much the best strategy. It is not possible to guarantee mutual acceptance between the anemone and clownfish of your choice if you buy them separately. Anemones of the Entacmaea *and* Macrodactyla *species seem to be widely accepted by many clownfish.*

RED SADDLEBACK ANEMONEFISH • *Amphiprion ephippiu*

This clownfish, also known as the tomato clown or fire clown, is slightly larger than the common clown (page 10) and lacks the vertical white banding found on some species. Its red-orange body is merely marked by an oval, dark brown patch two thirds along the flank. All the fins are plain red-orange in color. It has been reported that juveniles have a vertical thin white line passing down over their head, just behind the eye, but this fades with maturity.

Compatibility

While happy to be associated with the sea anemone, this species is a bold feeder and at times may become aggressive toward any other fish entering its territory.

Ungainly swimmers

The name "clownfish" has been applied to these fishes because of their waddling swimming action. In some literature they may even be called clown anemonefish.

The composition of the mucus covering the skin prevents the stinging cells (nematocysts) of the sea anemone's tentacles from "firing," thus providing the fish with immunity.

Size: 5.5 in (14 cm)

Origins

The Andaman and Nicobar Islands, Malaysia and Java.

Andaman and Nicobar Islands
Malaysia
Java

Fishkeeping data

Number per aquarium: Two to three, plus a suitable sea anemone.
Community/species tank: Community.
Swimming area: Middle and lower levels.
Food: All prepared foods, including freeze-dried and frozen, plus some vegetable matter.
Compatibility: Peaceful.
Availability: Regularly available (wild caught).
Captive breeding: No information.

BLACK-FOOTED CLOWNFISH

AMPHIPRION NIGRIPES

Although similar in appearance to Amphiprion ephippium *and* A. frenatus, A. nigripes *is much more subtly colored. It is a soft golden-brown, with a white stripe just behind the head. The pelvic fins are black, but the anal fin is not always so, hence the common name.* A. nigripes, *found off the Maldives Islands in the Indian Ocean, grows to 3.2 in (8 cm). In the wild it eats plankton and crustaceans, but finely chopped foods are ideal in the aquarium. Although a bold feeder,* A. nigripes *is a shy fish and best kept with other anemonefishes. It is an ideal subject to keep with invertebrates.*

Above: *The colors of* A. nigripes *are not as rich as some other clownfishes, and in this respect it is very similar to* A. perideraion.

PINK SKUNK CLOWNFISH • *Amphiprion perideraion*

As its popular name suggests, the body color of this clownfish is more subtle than the garish hues of the preceding species. A thin, vertical white stripe crosses the body at the rear of the gill cover and another white line runs from the top of the head right along the top of the dorsal ridge to the caudal peduncle. The pectoral, pelvic, and anal fins are the same color as the body, but the dorsal and caudal fins are almost colorless, with just a tinge of yellow. The skunk anemonefish, *Amphiprion akallopisos*, is similarly colored but lacks the white vertical stripe on the head.

Over its natural distribution range, it is usually found within the tentacles of the sea anemone *Heteractis magnifica.*

Sex reversal

The genus is also noted for its sex reversal characteristics. However, in this instance, all fish mature as males, only changing into a dominant female in order to head small colonies.

Origins

From Thailand to Samoa, including waters south of Japan down as far as the Great Barrier Reef and New Caledonia.

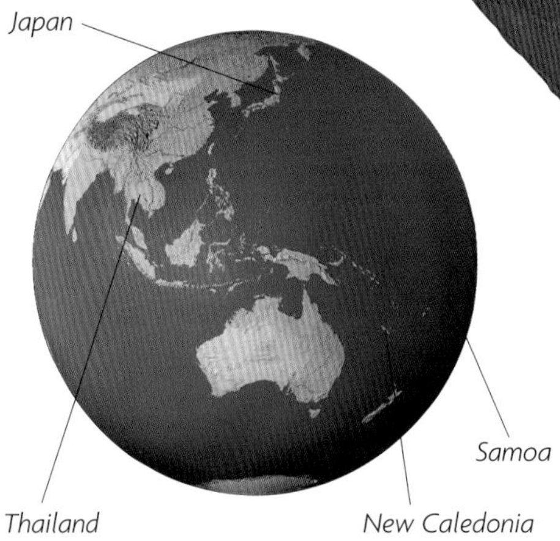

Japan

Samoa

Thailand

New Caledonia

Size: 4 in (10 cm)

Fishkeeping data

Number per aquarium: Two to three, plus a suitable sea anemone.
Community/species tank: Community.
Swimming area: Middle and lower levels.
Food: All prepared foods, including freeze-dried and frozen, plus some vegetable matter.
Compatibility: Peaceful.
Availability: Less frequently available (wild-caught).
Captive breeding: No information.

TOMATO CLOWNFISH

AMPHIPRION FRENATUS

The tomato clown, also known as the fire clown and the bridled clownfish, is very similar to A. ephippium, *but it has the white stripe behind the head (sometimes two in juveniles) and the body blotch is often larger. The confusion is not helped by the fact that some authorities call this fish* A. ephippium *or* A. melanopus. *Another name is* A. polylepis. *The tomato clown is found in the Pacific Ocean and grows to 3 in (7.5 cm). It is a bold feeder that accepts small crustaceans, small live foods, algae, and vegetable-based foods. It may be quarrelsome in confined spaces but perfectly compatible with invertebrates.*

Below: *Adult specimens of the tomato clown retain the vertical white stripe into adulthood. As with all fishes with similar features that are gathered over very wide areas by collectors not aware of other's efforts or descriptions, there is bound to be confusion with classification.*

MAROON CLOWNFISH • *Premnas biaculeatus*

This is one of the largest clownfish. In addition to this identifying factor, it also has two rearward-facing sharp spines on each cheek just below the eye. This feature accounts for its alternative common name of spinecheek anemonefish. The orange and red body color is more intense in the male fish and is crossed vertically by three narrow white bands. The female fish is much more likely to attain the maximum size and her body color can be much darker, almost black, with less distinct bands. These may almost disappear in some specimens.

Like all wild-caught anemonefishes, the maroon clownfish is susceptible to stress arising from poor handling and poor aquarium conditions. However, specimens obtained commercially from captive breeding programs are said to be much hardier.

It is usually found exclusively with the bubbletop sea anemone, *Entacmaea quadricolor*, although in the aquarium it may not be quite so dependent on a sea anemone as its relatives.

Origins

Western Indonesia as far north as Taiwan and south to the northern end of the Great Barrier Reef.

Its larger size, deep orange-red coloration and three white stripes makes the maroon clownfish easy to identify.

Fishkeeping data

Number per aquarium: One.
Community/species tank: Community.
Swimming area: Middle and lower levels.
Food: All prepared foods, including freeze-dried and frozen, plus some vegetable matter.
Compatibility: May bully.
Availability: Frequently available (wild caught).
Captive breeding: Possible.

Size: *6.6 in (17 cm), males are usually smaller*

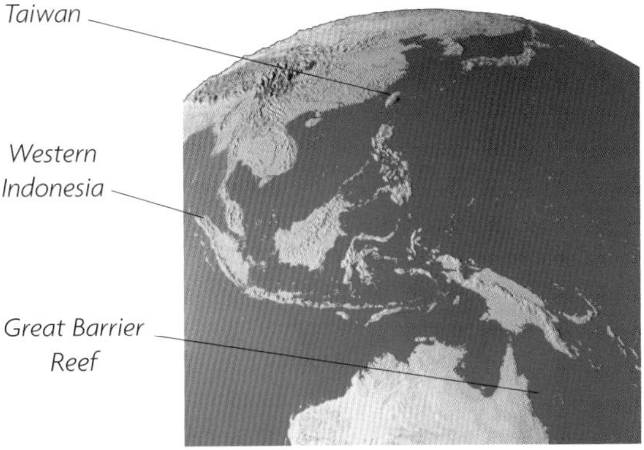

Taiwan

Western Indonesia

Great Barrier Reef

TWO-BANDED ANEMONEFISH

AMPHIPRION CLARKII

Also known as the banded clown and Clark's anemonefish, this species is highly variable in coloration depending on its location in the Indo-Pacific. Generally, the body is predominantly dark brown but for the ventral regions, which are yellow. All the fins, with the exception of the paler caudal fins, are bright yellow. Two tapering white bars divide the body into thirds; in juvenile forms, there is a third white bar across the rear of the body. A. clarkii grows to 4.7 in (12 cm) and enjoys a diet of small crustaceans, small live foods, algae, and vegetable-based foods. It is peaceful and makes an ideal choice for a mixed aquarium.

Above: *Clark's anemonefish can be found in many color variations and identification is therefore sometimes difficult. This bold fish is excellent for the community aquarium with little need for a host anemone in most cases.*

SERGEANT MAJOR • *Abudefduf saxatilis*

Look out from a glass-bottomed boat anywhere in the tropics and you are bound to see an oval, silvery fish with a number of vertical dark bands across its sides. In military terms, the sergeant major has an enormous parade ground, being found on rocky shores and coral reefs in both the Atlantic and Indo-Pacific areas, although to be accurate, there is more than one species with this common name. For example, *A. saxatilis* is from the Atlantic, while *A. vaigiensis* is found from the Red Sea eastward to Lord Howe Island in the Pacific, but *A. abdominalis* is restricted to the Hawaiian islands.

The body coloration is similar (give or take a blotch or two) from species to species. The body shape is very reminiscent of the freshwater sunfish, although the caudal peduncle is much shorter. Sergeant majors can be territorially aggressive.

Fishkeeping data

Number per aquarium: One.
Community/species tank: Community.
Swimming area: Middle and upper levels.
Food: All foods.
Compatibility: May bully.
Availability: Usually available (wild caught).
Captive breeding: No information.

Spacious aquariums can contain more fish from one species, providing they each have ample "territorial space."

Size: 6 in (15 cm)

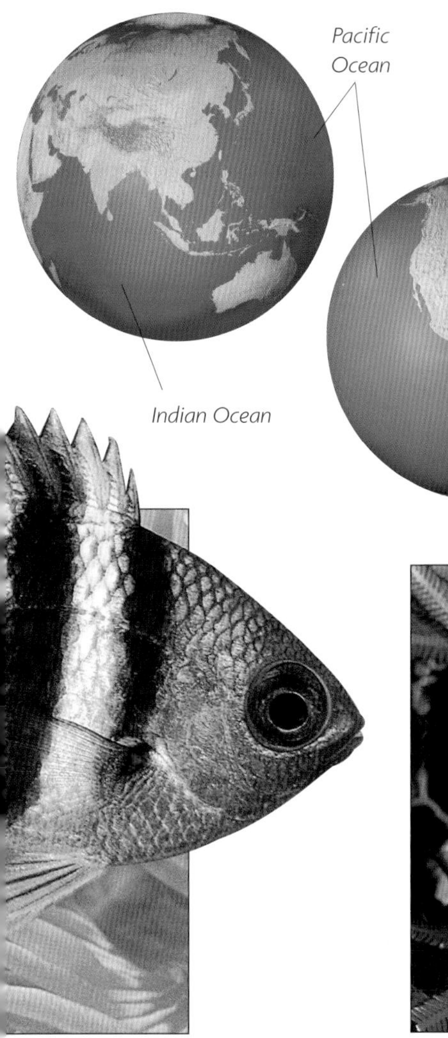

Origins

The Atlantic and Indo-Pacific areas.

BLUE VELVET DAMSELFISH
NEOGLYPHIDODON OXYODON

A vertical yellow stripe crosses the deep blue-black body just behind the head. The electric-blue wavy lines on the head and upper part of the body fade with age. This handsome but aggressive fish from the Pacific Ocean can be safely kept with invertebrates. It requires a diet of finely chopped meats, algae, and greenstuff, and grows to 5 in (13 cm). It may not adapt well if it is in less than first-class condition when you buy it.

YELLOWTAIL DEMOISELLE • *Chrysiptera parasema*

With the demoiselles we enter a veritable minefield of identification problems, because there are so many similar-looking, brilliantly blue-colored species from several different genera – without counting the blue-only juvenile forms. *C. parasema* has a touch more yellow to its posterior end than most, which should make identification slightly less problematical.

All damselfish are territorial and on the aggressive side. They will quarrel if there is not enough room (or too few hideaways) in the aquarium. Introducing lone specimens (often of any species of fish, not just the same species) into an aquarium already inhabited by damsels is asking for trouble. The best way to encourage them to settle in is to introduce a small number at once. Another trick is to relocate the aquarium decorations just before making any new introduction, so that everyone is too busy choosing new territories to pick a fight.

Electric blue, although a strikingly dazzling color, is common among several similar species of damselfishes.

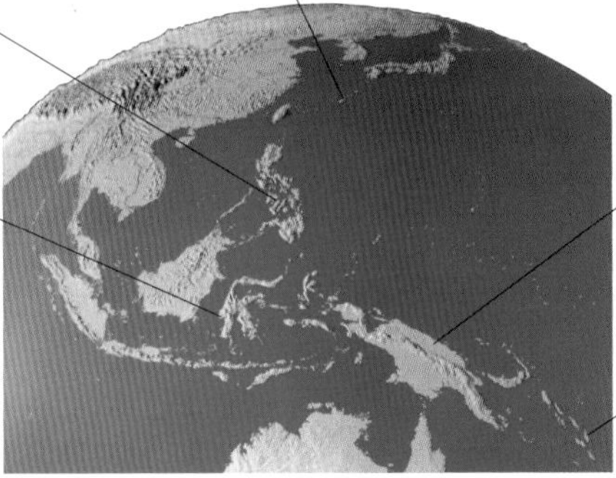

Ryuku Islands

Philippines

Sulawesi

Papua New Guinea

Solomon Islands

Origins

Philippines, Sulawesi, New Guinea, Solomon Islands, and Ryuku Islands.

Size: *2.75 in (7 cm)*

GREEN CHROMIS

CHROMIS VIRIDIS

This hardy, colorful shoaling species has a brilliant green-blue sheen to the scales. The caudal fin is more deeply forked than in some damselfishes. Keep these gregarious fishes in a shoal of at least six; individuals may go into decline in the aquarium. This lively and attractive species from the Indo-Pacific and Red Sea grows to 4 in (10 cm), is generally peaceful in the aquarium, and can safely be kept in a mixed fish and invertebrate setup. It is a shy feeder; offer it chopped meats.

Below: Chromis viridis *is a "cheap and cheerful" species and an ideal beginner fish that usually weathers early aquarium teething troubles better than most.*

Fishkeeping data

Number per aquarium: Two or three (or maybe two or three single specimens from different genera).
Community/species tank: Community.
Swimming area: Middle and upper levels.
Food: All foods.
Compatibility: Territorial, pugnacious.
Availability: Often available (wild caught).
Captive breeding: No information.

HUMBUG • *Dascyllus aruanus*

This fish's common name may intrigue anyone who is unfamiliar with the black-and-white striped peppermint confection of the same name.

The high-backed white body is crossed vertically by three black bands, which are united by a thick black edge along the top of the dorsal fin. The pelvic and anal fins are also black, but the pectoral and caudal fins remain clear. *Dascyllus melanurus* is often confused with this species. Although it lacks the black edge to the dorsal fin, it does have a black area across the caudal fin.

Damselfish congregate around coral heads and are only too quick to dash into the safety of the coral's branches. There is bound to be a wealth of such sanctuaries in this fish's natural range, from the Red Sea eastward to Lord Howe Island in the Pacific.

Fishkeeping data

Number per aquarium: Two to three (or maybe two to three single specimens from different genera).
Community/species tank: Community.
Swimming area: Middle and upper levels.
Food: All foods.
Compatibility: Peaceful.
Availability: Regularly available (wild caught).
Captive breeding: No information.

Domino, or three-spot damselfish

The three white spots on the jet black body of D. trimaculatus *are very reminiscent of a domino piece. A white spot appears on each side of the body just below the dorsal fin, with a third spot located on the dorsal surface just above the forehead. This coloration is limited to the juvenile form, as the adult fish is a sooty-grey, with less distinct (if at all visible) white markings.* D. trimaculatus *has a similar distribution to* D. aruanus *and the same requirements in captivity.*

A feature of the juvenile fish's behavior is its association with large sea anemones, a trait it loses when adult.

Size: *3.2 in (8 cm)*

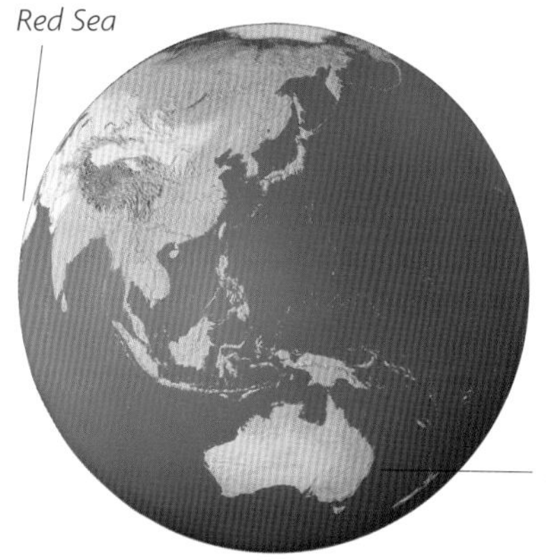

Origins

From the Red Sea eastward to Lord Howe Island in the Pacific.

Above: *As with all aggressive and territorial species, it is best to introduce* Dascyllus melanurus *to the aquarium last of all.*

BLACK-TAILED HUMBUG

DASCYLLUS MELANURUS

This fish is very similar to D. aruanus, *except that the black bars are more vertical and, as indicated by the common name, a black bar crosses the caudal fin. It is a shoaling species from the Western Pacific around the Philippines and Melanesia, and grows to 3 in (7.5 cm). In the aquarium the fish are aggressive and territorial and are best housed with other fish that can take care of themselves. They are bold feeders that will accept both chopped and dried foods, frozen mysis, and brine shrimp.*

Angelfishes

The Family Pomacanthidae contains many brilliantly colored fishes that range in size from very small to species that would feel more at home in a public aquarium. For example, members of the genus *Centropyge* can be accommodated quite easily in the home aquarium, whereas members of the genus *Pomacanthus* require much larger living quarters. A distinguishing physical feature of these fishes is the backward-projecting spine on the rear of the gill cover. Another characteristic of some members of this family is the complete difference in color patterning between juvenile and adult specimens.

Many angelfish are specific feeders and, unfortunately, it may not always be possible to acclimatize them to aquarium life. In the wild, some restrict their diets to sponges and similar nonreplicable foods, and it is difficult to provide a continuous supply of their required food.

Spawning usually occurs in the early evening, when pairs of fishes come together to release free-floating eggs at the climax of their courtship maneuvering. However, this is not likely to happen within the aquarium.

***Above:** The majestic angelfish,* Euxiphipops navarchus, *certainly lives up to its name. Suitable for the aquarium only as a young fish, when it is dark blue with vertical white stripes.*

BICOLOR ANGELFISH • Centropyge bicolor

As can be deduced from its distribution range, this species is a common occurrence. Its very distinctive patterning – forward half and caudal fin bright yellow, the remainder royal blue – makes identification simple. The blue area on the forehead helps to distinguish it from the very similar Coco's pygmy angelfish, *C. joculator*, which has a more swept-back forward edge to the blue area of coloration and orange-edged dorsal and anal fins. As it often feels insecure, the bicolor angelfish prefers areas of rocky rubble that offer welcome boltholes.

The coloration of this fish is not dissimilar to that of the much larger rock beauty, Holacanthus tricolor, *which lacks the dark patch above the eye.*

Cherubfish

The elongated oval body of C. argi *(3.2 in/8 cm) is a deep royal blue that extends into all the fins except the pectorals, which share the bright daffodil-yellow color of the head. The cherubfish frequents rocky rubble around the coral reefs of Bermuda, Florida, and the Caribbean. It feeds mostly on algal growths, so provide some vegetable matter within its diet.*

Size: 6 in (15 cm)

Origins

From Malaysia to Samoa, Japan to northwestern Australia.

Japan

Malaysia

Samoa

Northwestern Australia

Fishkeeping data

Number per aquarium: One or an established pair.
Community/species tank: Community.
Swimming area: Middle and lower levels.
Food: Appreciates some vegetable matter.
Compatibility: Peaceful.
Availability: Frequently available (wild caught).
Captive breeding: No information.

CORAL BEAUTY

CENTROPYGE BISPINOSUS

In young specimens, the head and body are outlined in deep purple; red flanks are vertically crossed by many thin purple lines. Adult fish, measuring 4.7 in (12 cm), have much larger areas of gold/yellow on the flanks, again crossed by dark vertical stripes. The pattern is very variable, however; for example, specimens from the Philippines have more purple and red coloration than those from Australasian waters. The fish will settle down in the aquarium if retreats are close at hand and are generally well behaved with invertebrates. They are grazers that appreciate meat foods and plenty of greenstuff in their diet.

Above: *This coral beauty shows the typical color pattern of young fish. Adult color varies depending on the location from which the fish have been collected.*

LEMONPEEL ANGELFISH • Centropyge flavissimus

The distinctive coloration of this species can lead to a little confusion between it and its relative, Herald's angelfish *(C. heraldi)*, but the latter lacks the outlining blue of the dorsal, anal, and caudal fins and the blue lips. However, just to make things difficult, a local population around the Cocos-Keeling and Christmas Islands in the Indian Ocean does lack the blue outline. Add to this the fact that the juvenile form of the mimic surgeonfish, *Acanthurus pyroferus*, which shares it distribution area (which also reaches as far as Easter Island), is also a plain yellow color, and you have even more opportunity for confusion.

Closer examination reveals that the lemonpeel hybridizes with the pearlscale angelfish *(C. vroliki)* and Eibl's angelfish *(C. eibli)*.

Origins

Cocos-Keeling and Christmas Islands; Ryukyu, Marianas to Rapa Island. Usually absent from Indo-Australian waters.

Size: 5.5 in (14 cm)

Fishkeeping data

Number per aquarium: One or an established pair.
Community/species tank: Community.
Swimming area: Middle and lower levels.
Food: Appreciates some vegetable matter.
Compatibility: Peaceful.
Availability: Frequently available (wild caught).
Captive breeding: No information.

***Below**: Look for the blue-edged fins if you are seeking this species, otherwise you might end up with something completely different, such as a juvenile surgeonfish.*

***Above:** Angelfish are spectacular fishes, but many are just too large for the average aquarium.* Centropyge *species, however, are ideally suited to the home aquarium.*

HERALD'S ANGELFISH

CENTROPYGE HERALDI

C. heraldi *is plain yellow, lacking even the blue details of* C. flavissimus. *Specimens collected from around Fiji have a black edge to the dorsal fin. In the world of marine fishes, monocoloration is quite exceptional. This makes the totally yellow Herald's angelfish from the Indo-Pacific easy to identify, and a popular choice. It grows to 4 in (10 cm) and is a peaceful fish, generally well behaved in the invertebrate aquarium. This grazing fish feeds mainly on greenstuff but also takes meaty foods.*

FLAME ANGELFISH • Centropyge loriculus

The coloration of the bright-orange body with its bright-yellow central area is further enhanced by vertical bands of deep blue-black. The edges of the soft-rayed portions of the dorsal and anal fins also have this blue-black coloration, but the caudal fin is red, shading to yellow. This species is somewhat shy and prefers to stay close to any convenient sanctuary-offering coral outcrop.

Male fish tend to be larger than females. All the fish start out as females, then males develop among a harem later.

Fishkeeping data

Number per aquarium: One or an established pair.
Community/species tank: Community.
Swimming area: Middle and lower levels.
Food: Appreciates some vegetable matter.
Compatibility: Peaceful.
Availability: Frequently available (wild caught).
Captive breeding: No information.

***Size:** 4 in (10 cm)*

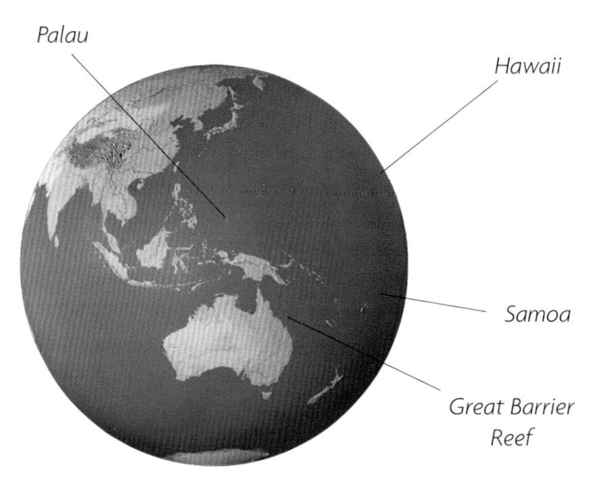

Origins

From Palau, just east of the Philippines, to Samoa and the Great Barrier Reef, although less frequent around Hawaii.

POTTER'S ANGELFISH

CENTROPYGE POTTERI

C. potteri *should not be regarded as an easy fish to keep; it demands excellent water quality and a varied diet. At first it may be a difficult feeder that needs tempting with live foods. However, once settled, it will usually accept the usual frozen marine food. Be sure to provide plenty of algae for the fish to graze on, as this makes up an important part of this species' diet.* C. potteri *is a peaceful fish that can often be kept with its own kind in pairs or in small groups in a sufficiently large aquarium. It is only found in the Hawaiian Islands. Measuring 4 in (10 cm), it falls somewhere in the middle size range between dwarf angels and the larger angels.*

Above: Centropyge *species are found at varying water depths. Potter's angelfish lives in 50 ft (15 m) of water.*

EMPEROR ANGELFISH • *Pomacanthus imperator*

This fish is a favorite (and very recognizable) species with fishkeepers. The body is traversed with diagonal, alternating stripes of blue and yellow, the snout is grey, the eye is hidden within a blue-edged dark stripe, and a yellow face mask is interrupted by a further dark area down across the gill cover. The anal fin is purple with red streaking, while the dorsal and caudal fins are yellow. It is reported that specimens from the Indian Ocean have rounded dorsal fins, not pointed as in other cases. However, all this gorgeous patterning is in the future because, like many of its relatives, the juvenile form is dark blue with white markings, in a vaguely concentric form in this particular species.

Juveniles tend to keep in or around boltholes and ledges in coral outcrops, but the adults are much bolder on the seaward coral faces of reefs that occur throughout its large distribution area.

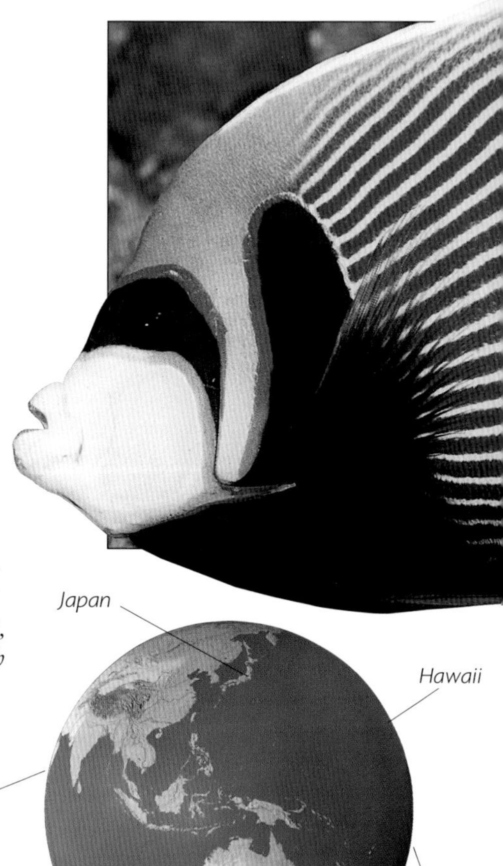

Origins

From the Red Sea to Hawaii, Japan to south of Tahiti.

Above: *Like many juvenile angelfish, the young emperor angelfish does not resemble the adult coloration form in any way.*

Size: 16 in (40 cm)

BLUE RING ANGELFISH

POMACANTHUS ANNULARIS

In adult blue ring angelfish, blue lines run from either side of the eye diagonally across the brown body. The lines rejoin at the top of the rear portion of the body. A dominant blue ring lies behind the gill cover. Juveniles are blue, with a distinctive pattern of almost straight transverse white lines. It is not hard to understand why juveniles and adults were once considered to belong to different species. Blue-ring angelfish, from the Indo-Pacific, grow to 16 in (40 cm) and are territorial. Although they are safe to keep with invertebrates when young, they become destructive with age. The fish are grazers that feed on meat foods and greenstuff.

Below: *If you have sufficient tank space, try keeping this reliable species.*

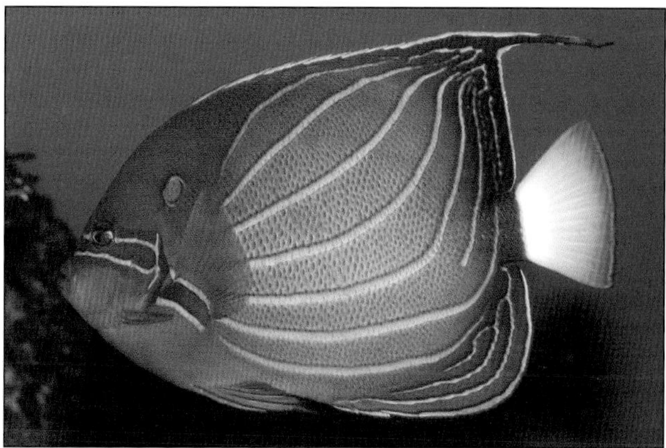

Fishkeeping data

Number per aquarium: One.
Community/species tank: Community.
Swimming area: All levels.
Food: Appreciates some vegetable matter.
Compatibility: Peaceful.
Availability: Frequently available (wild caught).
Captive breeding: No information.

FRENCH ANGELFISH • *Pomacanthus paru*

Most marine fishkeepers will be more familiar with the juvenile form of this species and its black body (and fins) crossed by vertical yellow stripes, as shown here. This is a far more interesting pattern than the adult fish's rather drab gray-black ground color with a faint dusting of yellowish flecks. Again, the adult fish is a rather large proposition to take on for the average domestic aquarium.

Fishkeeping data

Number per aquarium: One.
Community/species tank: Community.
Swimming area: All levels.
Food: Appreciates some vegetable matter.
Compatibility: Peaceful.
Availability: Frequently available (wild caught).
Captive breeding: Fry have been produced by using the "hand-stripping" method similar to that used for goldfish.

Breeding

French angelfish tend to pair bond in nature. The actual spawning process is a leisurely affair conducted in deeper wate with eggs being expelled and fertilized as the two fish swim upward throug the water. Reports indicat that breeding through "hand-stripping" adult fish of eggs and sperm (in a similar fashion to goldfish) is possible.

Size: *15 in (38 cm)*

Left: *This is a splendid example of an adult queen angelfish. Juveniles often act as "cleaners" to other fish visiting their locality.*

QUEEN ANGELFISH

HOLACANTHUS CILIARIS

This angelfish from the Western Atlantic looks very beautiful in the aquarium. Variations in color pattern occur and the body hues also appear to change under varying lighting conditions. Young fish have curving, royal blue vertical markings on a dark blue, gold-finned body. Adult coloration may also be adopted at half the fish's maximum size, given the proper diet and a spacious tank. H. ciliaris *grows large – up to 18 in (45 cm), grazing on meat foods and greenstuff. It is aggressive when young, intolerant of its own species, and can be very territorial. Although juveniles are safe with invertebrates, they become destructive with age. Queen Angelfish are reported to be prone to outbreaks of whitespot disease but can be successfully treated with copper-based remedies. They are quite resistant to such treatment and may even withstand some degree of overdosing without ill effects.*

Origins

A far-ranging species covering waters of the northern Gulf of Mexico to Brazil. There are even some reports of central Atlantic sightings around Ascension Island.

Butterflyfishes

Members of the Family Chaetodontidae are not dissimilar in shape or coloration to the angelfishes but lack the spine on the gill cover. Furthermore, feeding can present similar problems (see page 73). Most of the butterflyfishes have sharply pointed heads, often with elongated snouts that are ideal for picking among the coral for their food. They are typically active during daylight hours and retreat among the coral heads at night.

Although this group contains some of the most beautiful fishes to be seen around the coral reef, it is in their own best interests that they are left in their natural habitat. However, their expensive price tag will always mean that they are desired by the ambitious fishkeeper, who in all probability will be disappointed at their relatively short life in captivity, usually due to feeding problems. Because they are grazers of polyps and sponges, there is the added problem of keeping them in a reef tank containing live corals.

On the credit side, established pairs of butterflyfish generally do well, although their degree of "hardiness" is often due to their method of capture – net-caught specimens being far hardier than those captured by using drugs.

Above: *This young specimen of the elegant, boldly marked threadfin butterflyfish,* Chaetodon auriga, *has just a small extension to the rear of the dorsal fin from which the common name is derived.*

THREADFIN BUTTERFLYFISH • *Chaetodon auriga*

Many species of *Chaetodon* within the same distribution range are similarly colored, and the threadfin butterflyfish shares its "herring-bone" striped patterning of gray-black stripes on a silvery yellow background with two or three relatives. Additionally, the hiding of the eye in a vertical dark band is a common feature among butterflyfishes in general, serving to protect this vital organ from attack. It is especially effective when supplemented by a replica, or decoy, eye marking elsewhere on the body.

The common name refers to a threadlike extension to the rear of the yellow rear portion of the dorsal fin. While threadfins usually have an "eyespot" in the dorsal fin's yellow area, some populations on the seaward side of reefs do not. Normal foods include polyps, anemones, and algae.

Fishkeeping data

Number per aquarium: One.
Community/species tank: Community.
Swimming area: All levels.
Food: Polyps, anemones, and algae, but will take prepared foods.
Compatibility: Peaceful.
Availability: Frequently available (wild caught).
Captive breeding: No information.

Vagabond butterflyfish

Chaetodon vagabundus *may be distinguished from the threadfin by the black border running around the rear edge of the body and around the rear edges of the dorsal and anal fins. Two vertical black bars appear on the yellow caudal fin. The forehead is less steeply inclined. The diet follows that for the threadfin, and the distribution area is similar.*

Size: *9 in (23 cm)*

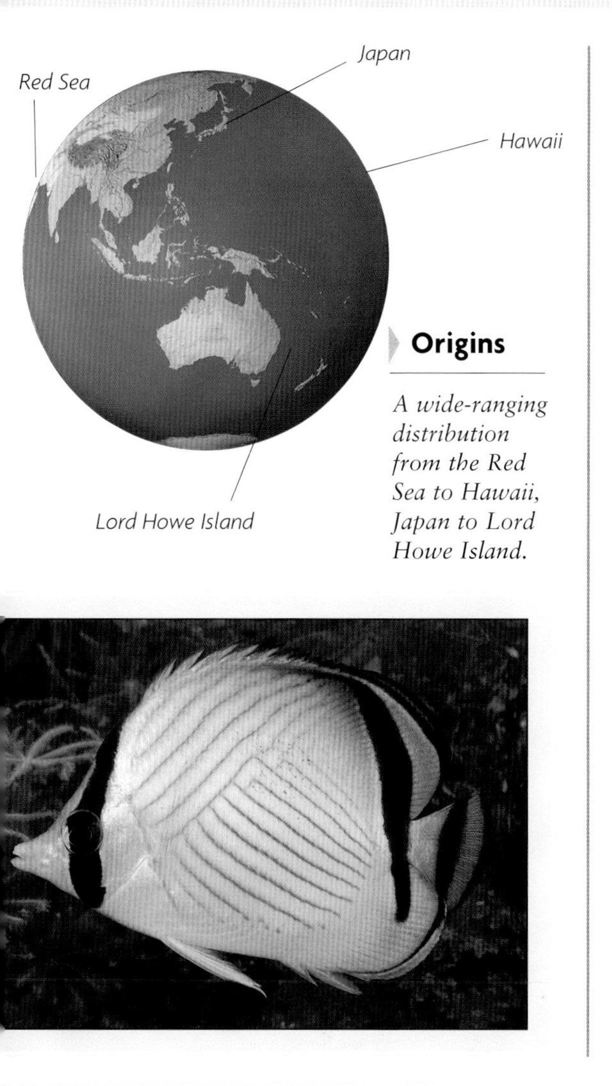

Origins

A wide-ranging distribution from the Red Sea to Hawaii, Japan to Lord Howe Island.

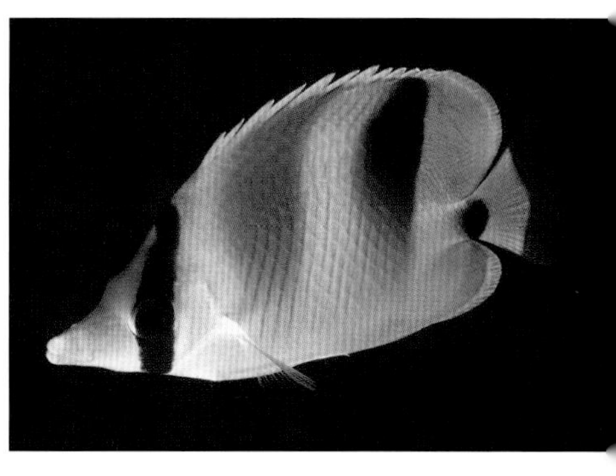

DOUBLE-SADDLED BUTTERFLYFISH

CHAETODON FALCULA

Two dark saddle markings cross the top of the body (developing on this young fish). The dorsal, anal, and caudal fins are yellow, and there is a black spot or bar on the caudal peduncle. The body and head are white. A vertical black bar runs down the side of the head and there are many thin vertical lines on the body. C. falcula, *found in the Indian Ocean, grows to 6 in (15 cm) and is often confused with* C. ulietensis, *which occurs in the wider-ranging Indo-Pacific waters, and has slightly lower reaching "saddles" and less yellow on top of the body and dorsal fin. Like most species in this family,* C. falcula, *which also goes by the common name of saddled butterflyfish, is not suitable for the beginner. It requires a diet of crustaceans, coral polyps, and algae. It is aggressive toward similar species and cannot be trusted with invertebrates.*

COPPERBAND BUTTERFLYFISH • *Chelmon rostratus*

While many butterflyfishes have almost identical coloration, namely bright yellow bars on a silvery background, the copperband is in a class of its own, except maybe for the very similar ocellate coralfish, *Parachaetodon ocellatus*, found further westward. However, its body shape is more rhomboid, while the copperband's is oblong.

The silvery body is crossed vertically by five black-edged orange bands that reach into the yellow-and-blue-edged pelvic, anal, and dorsal fins. There is a false eyespot in the rear portion of the dorsal fin. The main physical characteristic (apart from the coloration) is the long beaklike snout – ideal for reaching into coralline crevices for food.

The copperband may be slow to acclimatize to tank conditions and appears to do best in a well-established reef aquarium, where it can find enough to satisfy its coral-pecking instincts.

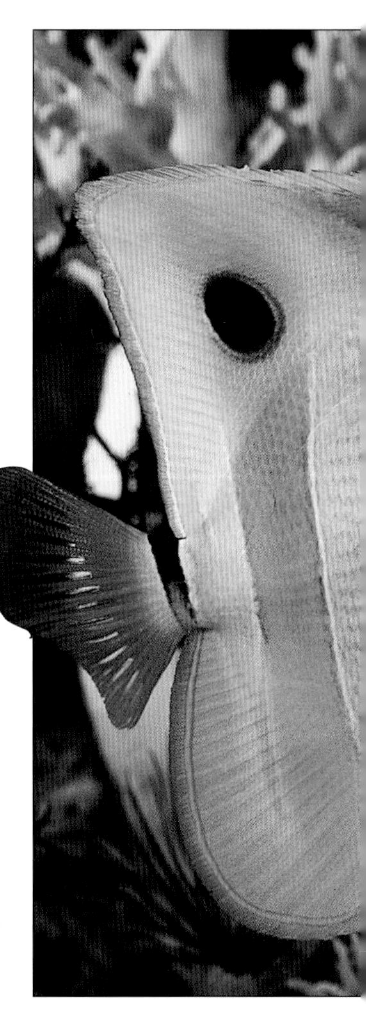

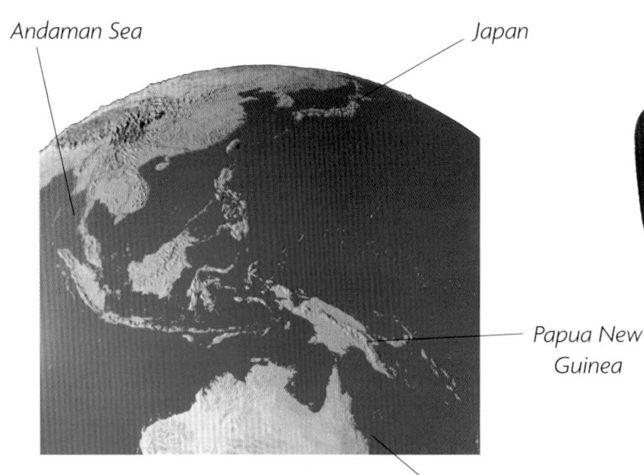

Origins

Found either as a solitary specimen or in pairs from the Andaman Sea, Papua New Guinea, Australia, and Great Barrier Reef north toward Japan.

Size: *8 in (20 cm)*

Fishkeeping data

Number per aquarium: One.
Community/species tank: Community.
Swimming area: Middle and lower levels.
Food: All "meaty" foods.
Compatibility: Peaceful.
Availability: Frequently available (wild caught).
Captive breeding: No information.

If you had to design a tool for collecting food from coral reef crevices, you would be hard pressed to come up with something better than this.

ADDIS BUTTERFLYFISH

CHAETODON SEMILARVATUS

The yellow body of Chaetodon semilarvatus *is crossed by thin orange vertical lines. A blue-black inverted teardrop patch covers the eye. Specimens from the Red Sea are difficult to obtain and therefore very expensive. This species, also known as lemonpeel butterflyfish and golden butterflyfish, grows to 8 in (20 cm), and grazes on crustaceans, coral polyps, and algae. The bright coloration and distinctive markings make it an attractive fish to look at, but in the aquarium it may be intolerant of other members of its own, or other, species and cannot be trusted with invertebrates. Nevertheless, it settles down readily in captivity.*

LONGNOSED BUTTERFLYFISH • *Forcipiger longirostris*

This species not only looks extremely similar to its relative, *F. flavissimus*, but also shares much of its distribution area. For those wishing to differentiate between the two species, *F. longirostris* has the longer snout and the steeper forehead, while *F. flavissimus* has a larger gape to its mouth.

The head coloration serves two purposes: the black upper half protects the eye (there is a small eyespot on the rear of the anal fin to divert attack away from the real eye), while the silver lower half effectively breaks up the fish profile, again perhaps to prevent detection from would-be predators. The whole body rearward of the gills, together with the pelvic, dorsal, and anal fins, is bright yellow, with bright blue on the edges of the single fins. The caudal fin is colorless.

Note that the real eye is hidden in a dark area with a "decoy" eye prominently displayed just below the caudal peduncle.

This is another instance of disruptive coloration, with bold areas of different color that break up the fish's outline.

Size: *8.5 in (22 cm)*

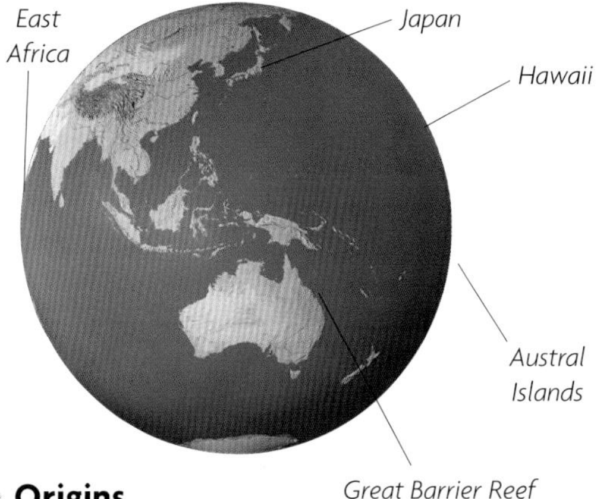

Origins

From East Africa to Hawaii, Japan to the Austral Islands, including the Great Barrier Reef.

Fishkeeping data

Number per aquarium: One.
Community/species tank: Community.
Swimming area: Middle and lower levels.
Food: All "meaty" foods.
Compatibility: Peaceful.
Availability: Frequently available (wild caught).
Captive breeding: No information.

WIMPLEFISH

HENIOCHUS ACUMINATUS

Two wide, forward-sloping black bands cross the white body. The rear parts of the dorsal, pectoral, and anal fins are yellow, while the pelvic fins are black. The front few dorsal fin rays are much extended. H. acuminatus, *from the Indo-Pacific and Red Sea, varies little in juvenile and adult coloration. Young specimens act as cleanerfishes. They grow to 7 in (18 cm), grazing on most frozen marine foods and greenstuff. Heniochus is easier to keep than the lookalike Moorish idol* (Zanclus canescens). *It is a companionable shoaling fish, but the natural leader in a group in captivity may develop into a bully. Do not keep these fish with invertebrates.*

Wrasses

The Family Labridae contains very many species of all shapes and sizes, although usually it is only juvenile members of the species that are suitable for the home aquarium. Again, there may be a vast difference both in color and patterning between juvenile and adult forms.

All wrasses appear to rest at night, often burying themselves in the substrate or spinning a cocoon of mucus to form a "sleeping bag." An aquarium furnished with many nooks and crannies is ideal for wrasses.

Perhaps the best-known species is the cleaner wrasse, *Labroides dimidiatus*, which feeds off parasites on other fishes. Unfortunately, this habit pushes the species into a "specific feeder" category, which means that in the average aquarium (containing few parasites) there may well not be enough food for this fish to thrive. For this reason ask your aquarium dealer for advice on buying this fish.

Many wrasses are brilliantly colored and would make attractive aquarium subjects. The larger species are not recommended for a reef-type tank, although smaller species are much more satisfactory in this respect.

Above: *The banana, or canary, wrasse* (Halichoeres chrysus) *is modestly sized, and a trio can be comfortably housed in the average-sized aquarium to form the basic "starting off" collection.*

CUBAN HOGFISH • *Bodianus pulchellus*

The stocky but streamlined body is bright crimson-red for most of its length; only the top half of the caudal peduncle is bright yellow. The pelvic, dorsal, and anal fins are also scarlet, while the caudal fin is mostly yellow; only the bottom edge carries the remainder of the red color from the body. The clear pectoral fins feature a black blotch at their tips. In some specimens, a lateral white stripe may be present.

This bold species acts as a cleanerfish when young and is often found in close association with the bluehead wrasse, *Thalassoma bifasciatum*.

Origins

Wide ranging across the Caribbean, from Florida to Brazil.

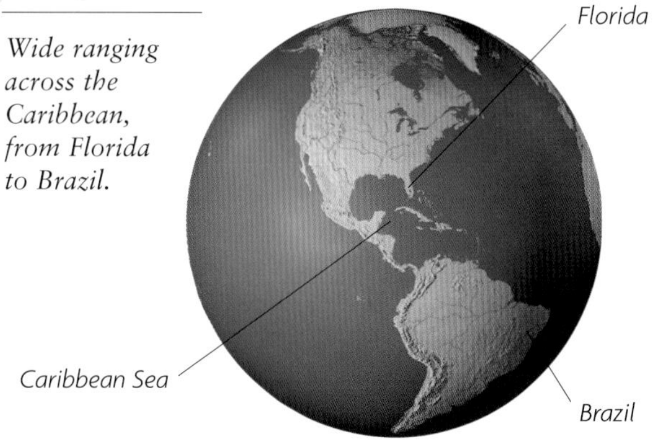

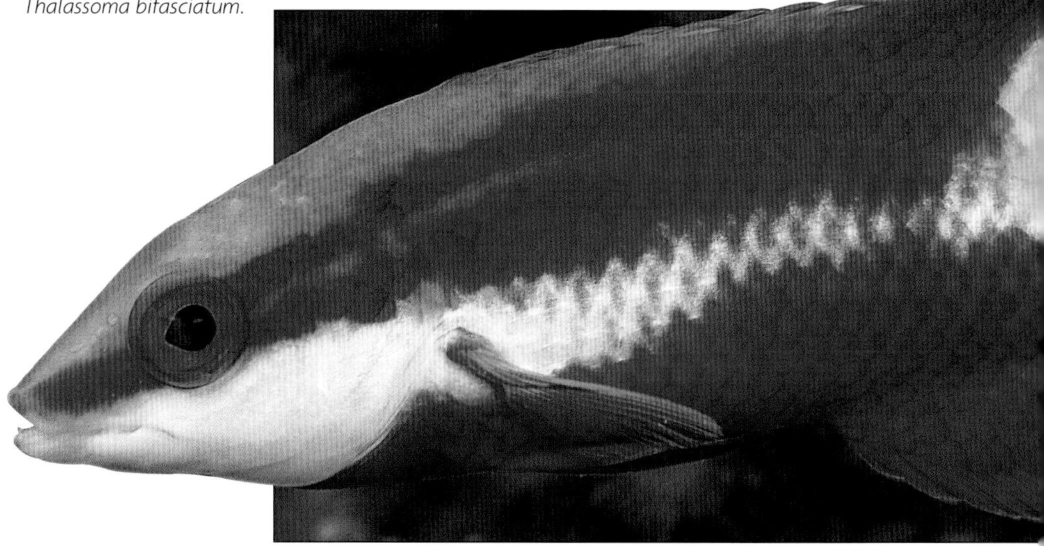

Size: *6 in (15 cm)*

Fishkeeping data

Number per aquarium: One.
Community/species tank: Community but not with invertebrates.
Swimming area: Middle and lower levels.
Food: All "meaty" foods.
Compatibility: Usually peaceful but intolerant of its own kind.
Availability: Frequently available (wild caught).
Captive breeding: No information.

SPANISH HOGFISH

BODIANUS RUFUS

Juvenile specimens are yellow with an area of blue along the upper body. Adult fishes show the standard red and yellow coloration, although the proportions may vary according to the habitat and depth of water. These brightly colored wrasses from the Western Atlantic grow to 8 in (20 cm), feeding boldly on crustaceans and shellfish meat at the bottom of the tank. They are peaceful community fish and safe to keep with invertebrates when young but become destructive with age and size. Like other members of the genus, juveniles perform cleaning actions on other fishes.

Above: *The change between juvenile (as shown here) and adult coloration is probably in order to adapt to the differing light levels in deeper water, where the adults eventually live.*

Below: *The spotfin hogfish takes its name from the black blotch on the end of each pectoral fin.*

HARLEQUIN TUSKFISH • Choerodon fasciatus

The coloration of this fish is quite stunning, with blue-edged bright orange bands crossing the blue-silver body vertically. The pelvic, dorsal, and anal fins are also orange-red with blue edges. The body area around the caudal peduncle is dark blue, and the caudal fin is white with a red final edge. The teeth are protrusible, ideal for moving rocks around in order to get at the invertebrate food beneath them, or for wrenching mollusks from rocks.

Fishkeeping data

Number per aquarium: One.
Community/species tank: Community but not with invertebrates.
Swimming area: Middle and lower levels.
Food: All "meaty" foods.
Compatibility: Usually peaceful but intolerant of its own kind.
Availability: Frequently available (wild caught).
Captive breeding: No information.

This adult tuskfish has lost the "eyespots" seen on the dorsal, anal, and pelvic fins of the juvenile form.

Size: *10 in (25 cm)*

Origins

Often found as solitary specimens in their native waters, ranging from Taiwan, Ryukyu Islands to Australia and the Great Barrier Reef.

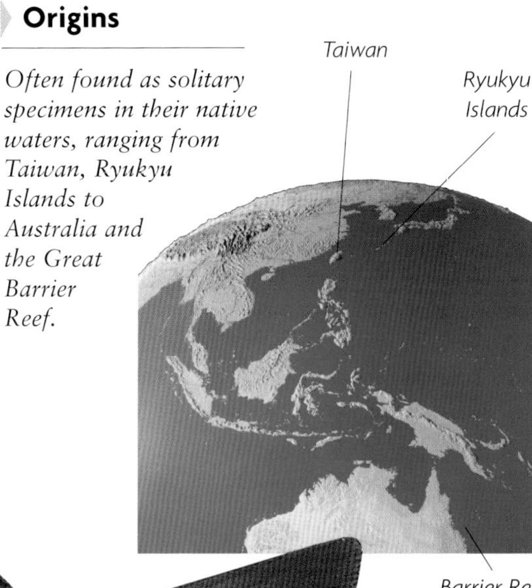

TWIN-SPOT WRASSE

CORIS AYGULA

A spectacularly colored species, both as a juvenile and as an adult. When young, this fish is white with two prominent orange spots on the dorsal surface. The front of the body, together with the fins, is covered with dark spots, and there are two white-edged black blotches on the dorsal fin. The adult is green with yellow-edged purple fins and is referred to as the Napoleon wrasse. This peaceful, bottom-feeding fish accepts small frozen marine animals and live foods. It can be safely kept with invertebrates when young but becomes destructive with age and size. It quickly grows to 8-12 in (20-30 cm).

Below: *Do not confuse the Napoleon wrasse with the Napoleonfish at three times its size!*

BIRDMOUTH WRASSE • Gomphosus coeruleus

The birdmouth wrasse is aptly named, for not only does it have an extended snout reminiscent of a bird's beak, but also a curious swimming action that gives the impression that the fish is "flying" through the water with a swooping action that is more avian than piscine.

The body of the male fish is green-blue, with a semicircle of yellow in the caudal fin. The dorsal and anal fins have yellow borders and the pectoral fins are black. The female is a dull brown color.

This constantly active species does not burrow into the substrate at night but hides up among the coral.

Origins

A wide natural distribution, ranging from the Red Sea to Hawaii, Ryukyu Islands to Lord Howe Island.

Red Sea
Ryukyu Islands
Hawaii
Lord Howe Island

Fishkeeping data

Number per aquarium: One.
Community/species tank: Community but not with invertebrates.
Food: All "meaty" foods.
Compatibility: Usually peaceful but intolerant of its own kind. Its constant activity may worry other less boisterous species.
Availability: Frequently available (wild caught).
Captive breeding: No information.

Size: 11 in (28 cm)

Above: *The legendary cleanerfish regularly services most fishes visiting its coral reef.*

CLEANER WRASSE

LABROIDES DIMIDIATUS

This is the most familiar of the wrasses because of its cleaning activities. This cleaning process, also practiced by some gobies and cleaner shrimps, is almost ritualistic. When approached by a cleaner wrasse, the subject fish – or host – often remains stationary with fins spread, in a head-up or head-down attitude. Sometimes the colors of the host fish fade, maybe so that the cleaner wrasse can see any parasites. The cleaner wrasse is found in the Indo-Pacific and grows to 4 in (10 cm). In the wild it feeds on the skin parasites of other fishes, but in the aquarium, finely chopped meat foods make an excellent substitute for this bold feeder. This constantly active but peaceful fish is safe with invertebrates.

Wolf in sheep's clothing

The elongate blue body of the cleaner wrasse has a horizontal dark stripe from snout to caudal fin. The mouth is terminal, and it is this feature that distinguishes Labroides dimidiatus *from the predatory lookalike* Aspidontus taeniatus, *the so-called false cleanerfish, which has an "underslung" mouth.*

BANANA WRASSE • Halichoeres chrysus

The brilliant yellow coloration, coupled with the shape of the body, should leave nobody in doubt as to the inspiration for this species' popular name. It has three black spots on the dorsal fin and one on the caudal peduncle. This helps to distinguish it from a similar-looking species, *H. leucoxanthus*, known as the canary-top wrasse, in which only the top half of the body is yellow; the lower half is silvery-white.

Origins

From Christmas Island to Marshall Islands, south Japan to southeastern Australia.

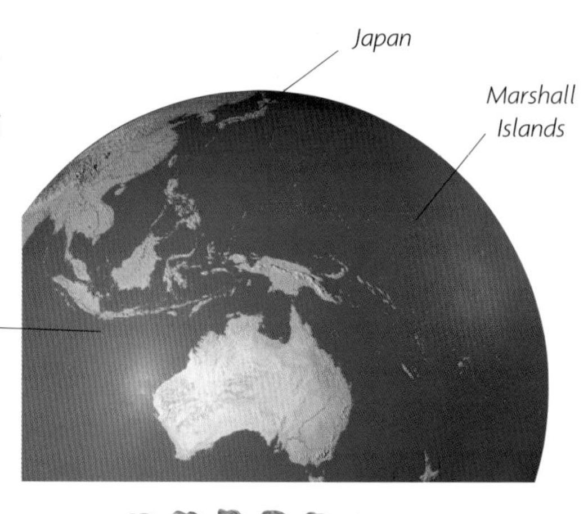

Fishkeeping data

Number per aquarium: Two to three.
Community/species tank: Community but not with invertebrates.
Swimming area: Middle and lower levels.
Food: All "meaty" foods.
Compatibility: Usually peaceful but may quarrel among themselves.
Availability: Frequently available (wild caught).
Captive breeding: No information.

Size: 4.7 in (12 cm)

Above: *However you look at it, the sixline wrasse is a smart character, from its horizontally striped eye to the tip of its green tail fin.*

SIXLINE WRASSE

PSEUDOCHEILINUS HEXATAENIA

The body shape of this species from the Indo-Pacific is one of classic symmetry, with the dorsal and ventral profiles mirroring each other perfectly. The coloration consists of horizontal stripes of alternating yellow and purple. The dorsal and caudal fins are yellow, while the pelvic and anal fins are purple. There is a white-edged black spot on the caudal peduncle. Despite the fish's attractive features, it tends to shy away from exhibiting them, preferring to frequent the coral branches rather than more open water. It grows to 3 in (7.5 cm) and readily accepts most marine frozen, live, and flake foods. Keep them in a community tank, where they constantly inspect the rockwork for titbits and, maybe, worry the snails. They are generally peaceful, except with the same or similar species. The sixline wrasse is an interesting, inexpensive, and disease-resistant species that can be heartily recommended to any newcomer to the hobby.

Think ahead

Many of the wrasse family are attractive fishes in their juvenile forms but become quite large, outgrowing their living quarters and losing their colors as they mature. This is something to bear in mind when selecting wrasses for your collection.

Surgeonfishes & tangs

Members of the Family Acanthuridae have oval or disc-shaped bodies, edged with long-based dorsal and anal fins. The head has a steeply rising forehead with the eyes mounted high up. The aptly named surgeonfishes carry an erectile scalpel on each side of the caudal peduncle, which can be used either offensively or defensively.

Members of this group are usually active during daylight hours and mostly herbivorous, requiring a good proportion of vegetable matter in their diet. Most are intolerant of their own kind, especially those of similar colors or shapes, but in a large aquarium, it is possible that oval-shaped fish will co-exist with the disc-shaped species without constant quarreling.

The popular name of "tang" is likely to be an abbreviation of the German "seetang," meaning seaweed. This reflects the fishes' fondness for consuming large quantities of algae and kelp as its natural diet. In addition to offering such a diet, it is also important to provide a flow of well-oxygenated water by means of a powerful filtration system with, perhaps, extra aeration.

Above: *What better common name could there be for these fishes than "surgeonfish." The "scalpel" near the junction of the tail and body shines white here as a warning to would-be predators to keep away.*

POWDER BLUE TANG • Acanthurus leucosternon

The steeply rising forehead and oval body are typical of the genus. Furthermore, the black head, white throat, and blue body surrounded by yellow dorsal and anal fins, plus a black-decorated caudal fin make recognition of this species easy. The "scalpel" is disguised in the yellow background of the caudal peduncle. The small, pointed mouth, well-equipped with teeth, is set well forward for efficient grazing on algae or capturing planktonic food. Although the fish often occur in large colonies in nature, most aquatic dealers separate young specimens from one another as they have a penchant for squabbling. Powder blues (and the related powder brown, A. japonicus) have a wide distribution area.

When buying surgeons and tangs, look out for "pinched-in" bodies, especially in the head region. These fishes are constantly eating in the wild and need adequate, vegetable-based diets.

Despite the fish's overall "blueness," the scalpel is visible against its bright yellow border.

Fishkeeping data

Number per aquarium: One.
Community/species tank: Community.
Swimming area: All levels.
Food: Will accept most foods, including "meaty" types, but benefits mostly from a high vegetable matter content in its diet. Forages for food continuously.
Compatibility: Quarrelsome with its own kind but usually ignores invertebrates.
Availability: Usually available (wild caught).
Captive breeding: No information.

Size: 9 in (23 cm)

Origins

From East Africa to southwest Indonesia.

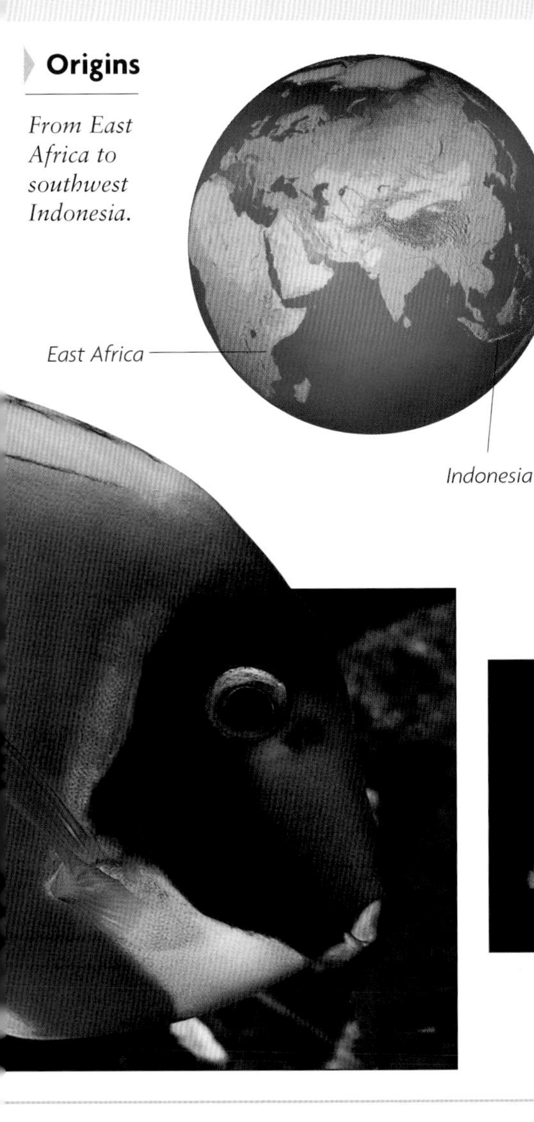

BLUE TANG

ACANTHURUS COERULEUS

Young fishes are yellow with blue markings around the eye – do not confuse them with Zebrasoma flavescens *(see page 142). Small specimens can be quite aggressive and may become bullies if established in the aquarium before other fishes, but this tendency generally decreases with time. Do not keep blue tangs with invertebrates. As they age, the fish develop narrow blue lines, the adult fish being darker blue than the "almost adult" fish. The scalpels on the caudal peduncle are ringed with yellow or white in mature fishes. Its dark blue coloration defies the bright yellow of the juvenile. Blue tangs are found in the Western Atlantic and grow to 6 in (15 cm). They feed mainly on algae.*

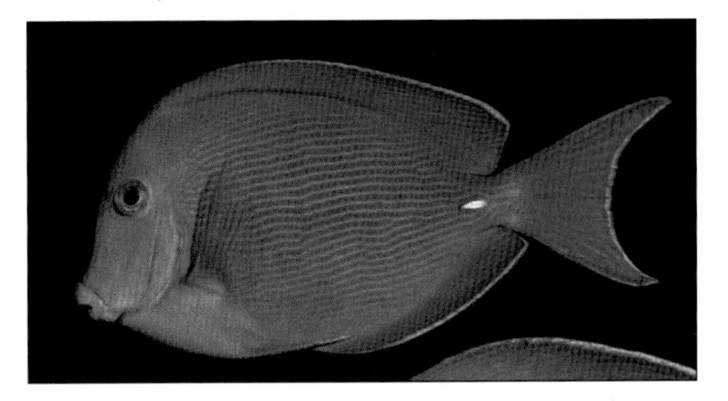

Above: *The scalpels of surgeonfishes are usually well hidden, but the color scheme of this species gives the game away.*

ORANGE-SPINE UNICORNFISH • *Naso lituratus*

Although this species is technically classified as a surgeonfish in most literature, it is also known as a member of the unicornfish group (Nasinae), some of which have a single hornlike projection from the forehead. Additionally, the "scalpels" are permanently erected and are arranged as a pair of bony projections on each side of the caudal peduncle.

The body coloration is olive-brown, while the whitish face has black markings and bright red lips that account for the fish's popular name of lipstick tang. The long-based dorsal fin is bright yellow and the almost equally long anal fin is brown. There is a black border around the caudal fin. As may be surmised by its size, it is an ocean-going species and naturally requires adequate aquarium space to contain its constant movements.

Origins

From the Red Sea to Hawaii, Japan to the Great Barrier Reef.

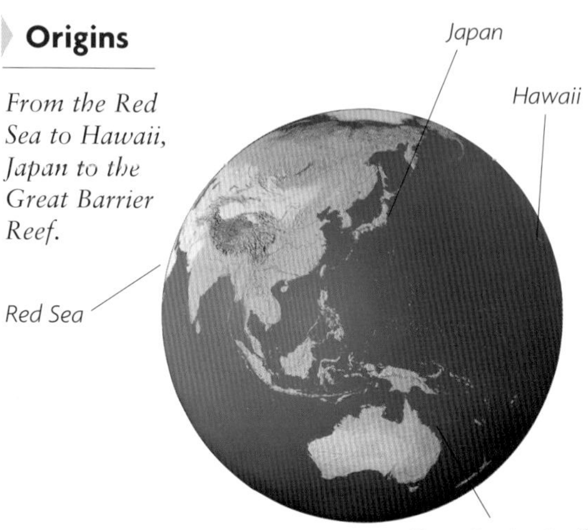

Fishkeeping data

Number per aquarium: One.
Community/species tank: Community.
Swimming area: All levels.
Food: Will accept most foods, including "meaty" types, but benefits mostly from a high-vegetable matter content in its diet. Forages for food continuously.
Compatibility: Quarrelsome with its own kind but usually ignores invertebrates.
Availability: Usually available (wild caught).
Captive breeding: No information.

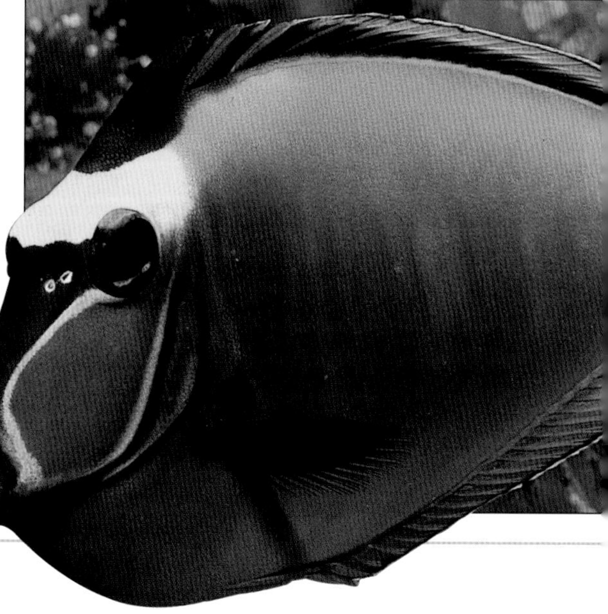

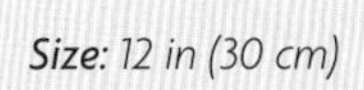

Size: *12 in (30 cm)*

Left: *The scalpels on this species are not as large as their disguising color panel.*

ACHILLES TANG

ACANTHURUS ACHILLES

The brown, oval body is offset by yellow-red baselines to the dorsal and anal fins. The white marking on the gill cover behind the eye and the dull white patch on the chest are shared by other members of the family, but the feature that positively identifies this fish is the teardrop-shaped orange-red area on the caudal peduncle, in which the scalpels are set. Young fish do not have nearly as many red markings as adults. They grow to 10 in (25 cm). The Achilles tang is normally peaceful but very delicate. It is compatible with most fish but may fight at first with other members of its own family. Do not add it to the tank until the first fish are established and do not keep it with invertebrates. The fish are found in the Pacific and graze on growths of algae (the high forehead makes this easy). In the aquarium they are shy grazers that will accept the usual protein foods, such as gamma-irradiated frozen foods and live brine shrimp, plus algae and other greenstuff.

REGAL TANG • *Paracanthurus hepatus*

This aquarium favorite is easily identified by its royal blue body, decorated with the familiar painter's palette outline, and the bright yellow caudal and pectoral fins.

All surgeons and tangs are very dependent on well-oxygenated water and are also warmth loving; a temperature of 79-82°F (26-28°C) suits them fine. Keep the specific gravity as near to the level they would experience in nature and also ensure that the pH is consistently stable at about 8.3.

Like all members of this family, they can be territorially minded, so adequate tank space with plenty of retreats is called for.

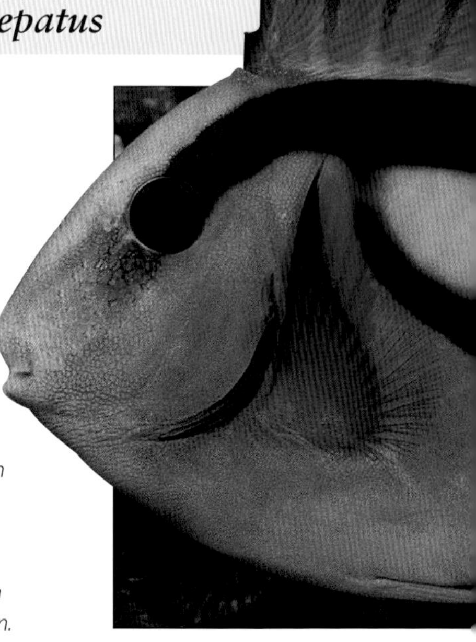

The "palette" pattern on the body is instantly recognizable, but the dorsal fin pattern and bright yellow caudal fin also deserve appreciation.

Origins

A wide-ranging species, from East Africa to the mid-Pacific Line Islands, southern Japan to the Great Barrier Reef, New Caledonia, and Samoa.

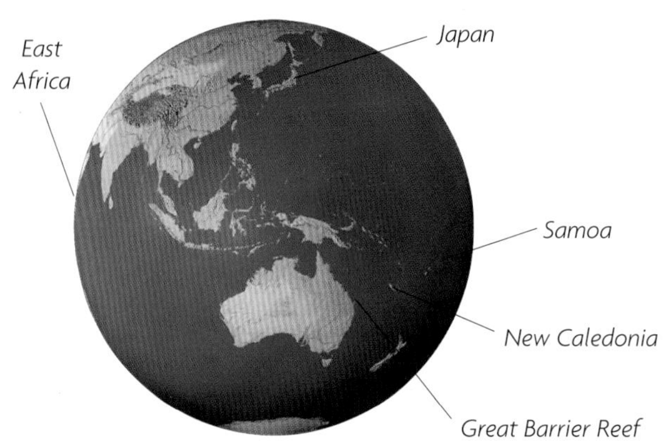

Fishkeeping data

Number per aquarium: One.
Community/species tank: Community.
Swimming area: All levels.
Food: Will accept most foods, including "meaty" types, but benefits mostly from a high-vegetable matter content in its diet. Forages for food continuously.
Compatibility: Quarrelsome with its own kind but usually ignores invertebrates.
Availability: Usually available (wild caught).
Captive breeding: No information.

***Size:** 12.25 in (31 cm)*

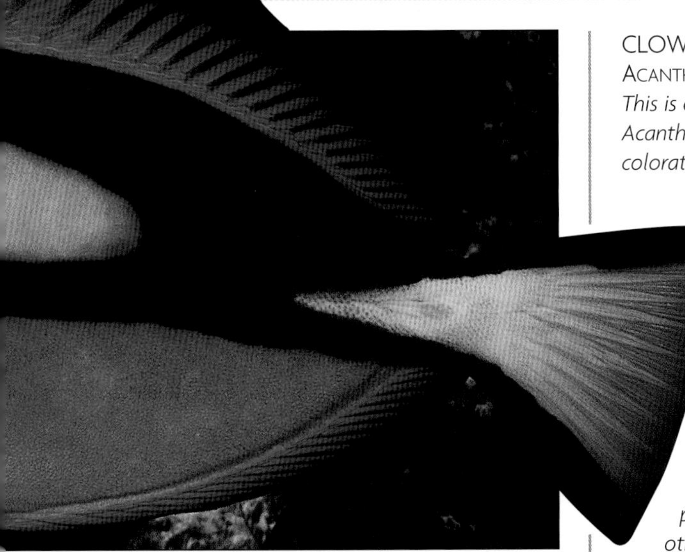

CLOWN SURGEONFISH

ACANTHURUS LINEATUS

This is one of the species of the Acanthuridae family with a split level of coloration: there is a lighter area to the lower body, with dark-edged, parallel longitudinal lines above. The pelvic fins are yellow. This striped pattern accounts for the fish's other common names – blue-lined surgeonfish and pyjama tang. Ideally, keep only one of these fish in an aquarium (unless you have a very large tank with space for several), as small specimens, in particular, can be quarrelsome. Like other surgeonfishes, A. lineatus *appreciates some coral or rockwork to provide welcome shelter. The fish are found in the Indo-Pacific and require a diet of algae. They will grow to 6 in (15 cm).*

***Left:** The colorful patterning of the clown surgeonfish looks quite stunning under aquarium lighting but may appear very different when viewed "down in the depths."*

YELLOW TANG • *Zebrasoma flavescens*

This brightly colored yellow fish has a high profile and stands out well in the aquarium. It relishes vegetable matter and will happily munch on scalded lettuce leaves. These can be conveniently secured between the jaws of a magnetic algae scraper to anchor them into an "easy-eating" position.

This species cannot be confused with other all-yellow fishes, such as the lemonpeel angelfish *(Centropyge flavissimus)*, Herald's angelfish *(C. heraldi)*, or even the related juvenile mimic surgeonfish *(Acanthurus pyroferus)*, because of its distinguishing tall finnage and more laterally compressed body.

Fishkeeping data

Number per aquarium: One.
Community/species tank: Community.
Swimming area: Middle and lower levels.
Food: Will accept most foods, including "meaty" types, but benefits mostly from a high-vegetable matter content in its diet. Forages for food continuously.
Compatibility: Quarrelsome with its own kind but usually ignores invertebrates.
Availability: Usually available (wild caught).
Captive breeding: No information.

Size: *8 in (20 cm)*

Origins

From Ryukyu Islands eastward, although most frequently found around Hawaii.

This is a fine specimen of yellow tang. Avoid tangs with concave bodies, especially in the head and dorsal areas.

SAILFIN TANG

ZEBRASOMA VELIFERUM

At first glance this species, from the Indo-Pacific, is reminiscent of the freshwater discus, Symphysodon *sp., but a closer look will reveal that the actual body shape is not round but a tapering, forward-sloping oval. It is the large, wide dorsal and anal fins that add the illusory "roundness." It grows to 16 in (40 cm).*

The pale cream background color is crossed vertically by five or six brown bars, the darkest two of which are found at the front, crossing the eye and the gill cover. Some bars also carry gold-red streaking along their lengths. The fins repeat the alternate brown-and-cream patterning in what appears as almost concentric circles around the body. The caudal peduncle is light blue with a dark blue scalpel; the caudal fin is yellow. Juveniles (shown here), which are yellow with alternating dark bars, frequent shallower waters than adults.

The fish appreciate a diet with a high vegetable content, as well as "meaty" foods. They may be quarrelsome with their own kind but usually ignore invertebrates.

Algae eaters

Keeping a tang in your aquarium is an ideal way of keeping unwanted growths of algae under control. However, some macro-algae is beneficial, as it removes nitrates and adds trace elements.

Triggerfishes

Members of the Family Balistidae have deep, triangular, tapering bodies, with only rudimentary stumps where the pelvic fins would normally emerge. However, their main characteristic is the locking action found on the first of the two dorsal fins. The second spine of the first dorsal fin locks it up into a vertical position, which can effectively lodge the fish into a crevice or, alternatively, prevent it being swallowed by another larger fish. (Remember that it may also catch in the fishkeeper's net!) The first dorsal fin can only be unlocked by releasing the "trigger," thus allowing the fin to settle back into a groove on the back of the fish, where it is normally carried flat and out of sight until needed.

Do not keep these fish with invertebrates on which they prey in nature. In captivity, they are often quite quarrelsome among themselves, so are best kept as single specimens. They have a rather unusual swimming action, propelling themselves through the water by using a side-to-side action of their dorsal and anal fins, rather than their tails. In nature, breeding occurs on the seabed, where eggs are deposited in shallow pits dug out by the male fish.

Above: *In this clown triggerfish,* Balistoides conspicillum, *the first dorsal fin has been locked in an upright position by the second spine, aptly called the "trigger." Normally, this fin is carried flat against the body.*

CLOWN TRIGGERFISH • *Balistoides conspicillum*

Like many marine species, the clown triggerfish has very disruptive color patterning. The lower half of the dark body is covered with large white ovals, while the upper body immediately beneath the dorsal fin has a bright-yellow "saddle." The eye is hidden, but a bright-yellow stripe crosses the snout below it. The mouth is accentuated by a bright-yellow surround and an additional white line beyond it. The teeth are very sharp, so handle this fish with care. Although triggerfishes have two separate dorsal fins, the first fin is normally folded flat and carried in a groove immediately above the yellow saddle marking.

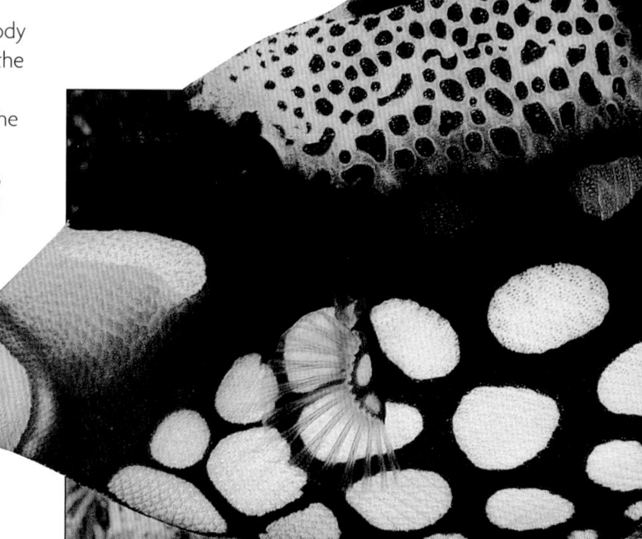

Fishkeeping data

Number per aquarium: One.
Community/species tank: Community, with larger fishes but not invertebrates. Often best kept as individuals in a species aquarium.
Swimming area: Middle and lower levels.
Food: All "meaty" foods.
Compatibility: Territorial.
Availability: Frequently available (wild caught).
Captive breeding: No information.

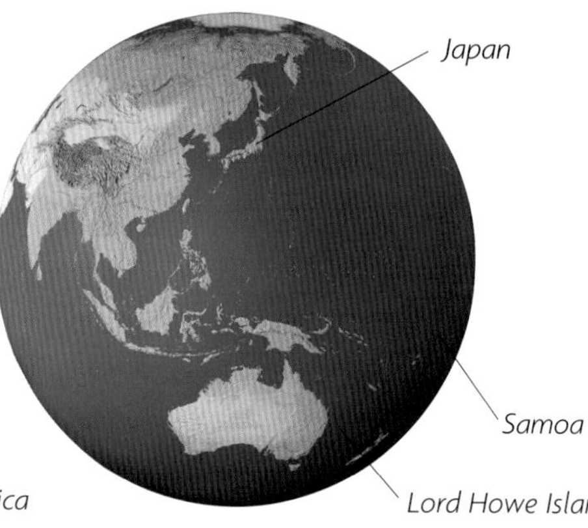

Origins

The clown triggerfish has a wide natural distribution, frequenting waters as far apart as East Africa to Samoa, Japan to Lord Howe Island.

Size: *20 in (50 cm)*

NIGER TRIGGERFISH

ODONUS NIGER

The body is a uniform dark blue-green, with the dark-centred scales creating a reticulated effect. The head region may be a shade or two lighter. Some facial markings may be visible, running from the snout up toward the eye. As the common name implies, the teeth are bright red. The outer margins of the fins are a bright blue, and the caudal fin is often very well produced, giving the fish a lyretail appearance. In the wild, when frightened, this is the only part of the fish to be seen, as it dashes into a coral crevice for shelter, leaving its tail sticking out. It is generally considered to be a peaceful species in the aquarium, but do not keep it with invertebrates. It accepts all "meaty" foods and grows to 16 in (40 cm).

Above: *Note the "waviness" of the rear dorsal and anal fins; these are providing propulsion rather than the caudal fin.*

Right: *The swimming action of triggerfishes is different to that of most fish. They wave their dorsal and anal fins from side to side, rather than use their caudal fin.*

PICASSOFISH • Rhinecanthus aculeatus

With a native common name – Humuhumu – as striking as its color patterning, this species always invites attention. Although the mouth is relatively small, the yellow lips and elongated yellow line that runs horizontally rearward across the gill cover give the impression of an immense gape – maybe a defensive coloration? The top of the head is yellowish, the throat and ventral area white. The lower half of the body from the pectoral fins to the caudal peduncle is decorated with diagonal alternate stripes of white and black. The caudal peduncle itself has horizontal black-and-white lines. A blue-edged dark stripe straddles the eye and covers the rear edge of the gill cover, while the upper half of the body is creamy white. The dorsal surface is separated from the diagonal stripes by brown-gray smudges through which brown stripes rise to meet the rearward of the two dorsal fins.

Despite appearances, the Picassofish's mouth is quite small. Note this fish's erect dorsal fin, which is not often seen.

Fishkeeping data

Number per aquarium: One.
Community/species tank: Community, with larger fishes but not invertebrates. Often best as individuals in a species tank.
Swimming area: Middle and lower levels.
Food: All "meaty" foods
Compatibility: Territorial.
Availability: Frequently available (wild caught).
Captive breeding: No information.

Size: *10 in (25 cm)*

Origins

Indo-Pacific waters from East Africa to Hawaii, Japan to Lord Howe Island.

Below: *Triggerfish spawnings may be feasible if their aquarium is sufficiently large.*

QUEEN TRIGGERFISH

BALISTES VETULA

Dark lines radiate from around the eyes and there are striking blue facial markings. The tips of the dorsal fin and caudal fins become filamentous with age, especially in the male, which is larger and more colorful than the female. This beautifully marked species may become hand tame in captivity. However, take care when hand feeding, as the fishes' slightly obtrusive teeth are very sharp. Try impaling pieces of food on a cocktail stick before offering them to the fish. They are bold feeders, accepting good-sized pieces of smaller frozen fishes, crustaceans, and mollusks, plus the usual frozen foods. B. vetula *is found in the tropical western Atlantic and grows to 10 in (25 cm). Do not keep it with small fishes or invertebrates. Although peaceful with other species, it will quarrel with its own kind.*

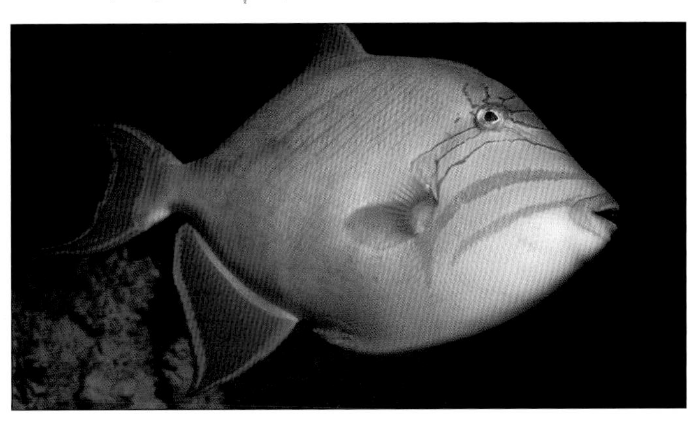

Groupers & basslets

The Family Serranidae is very large, containing both brilliantly colored, modest-sized specimens, and drab-colored monsters of the deep. Body shape tends to be stockily cylindrical, with spiny dorsal fins.

Many members are bottom-loving fish that constantly patrol the rubble at the base of coral reefs, while others may lie in wait, ready to pounce on any passing "meal." Sex reversal is common, with a female likely to change into a male as the need occurs, say, should the dominant male of a group of fishes cease to fulfill his function for whatever reason.

With the larger specimens, the production of large amounts of waste material demands an efficient filtration system and more frequent partial water changes to keep water conditions at their optimum.

The smaller specimens are especially suitable for the home aquarium. Such species include the royal gramma and its look-alike royal dottyback, both of which make excellent and highly colorful reef tank inhabitants. While they are included within this section for convenience, they are not strictly serranids but are actually members of the Grammidae and Pseudochromidae families, respectively.

Above: *Huge shoals of these wreckfish* (Pseudanthias squamipinnis) *are a common sight in their native habitat. Most will be females in a "harem" with their attendant solitary, dominant male.*

PANTHERFISH • *Chromileptis altivelis*

Despite its large size, there is something very graceful about this species as it cruises slowly around its necessarily large aquarium. The creamy-white body and fins are covered with black spots, which give rise to the alternative popular name of polkadot grouper. Yet another name, the humpbacked grouper, is inspired by the smallness of the head, which accentuates the height of the dorsal profile. Juvenile forms have larger black blotches that become smaller but more numerous with increasing adulthood.

In its natural home, this species is a valued food fish. Despite its apparently small mouth, it would be prudent not to keep it with small fishes, just to be on the safe side.

The large fins are used to control this fish's slow and graceful movement around the aquarium.

Fishkeeping data

Number per aquarium: One.
Community/species tank: Community.
Swimming area: All levels.
Food: Prepared foods, including dried and frozen products. Live foods (goldfish, bait fish, etc.) might give them ideas above their station with regard to other tankmates.
Compatibility: Reasonably well-mannered with other suitably sized fishes.
Availability: Frequently available (wild caught).
Captive breeding: No information.

Size: 27 in (70 cm)

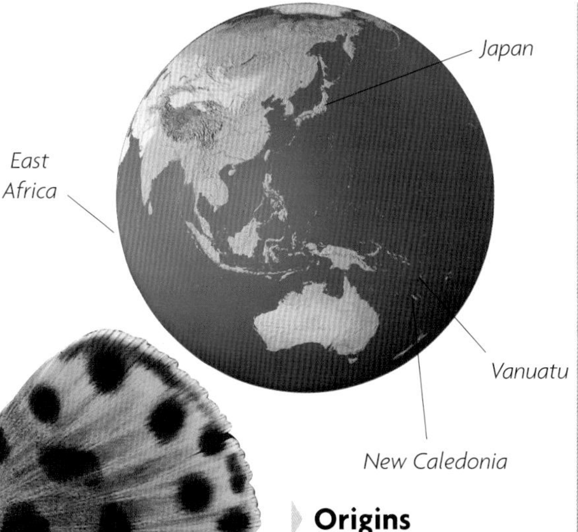

Origins

From East Africa to Vanuatu, southern Japan to New Caledonia.

MARINE BETTA

CALLOPLESIOPS ALTIVELIS

When looking at this very beautiful and deceptive fish, the trick is to decide which way it is facing, since the dorsal fin has a "false-eye" marking near its rear edge. The dark brown body is covered with light blue spots and the very elongated fins are similar to those of the freshwater Siamese fighting fish, Betta splendens, *hence the common name. The tail allegedly resembles the head of the moray eel – a useful defence against predators. This species from the Indo-Pacific spends much of its time in a "head-down" hunting position. Offer it a diet of frozen smaller fishes and meaty foods. It grows to 6 in (15 cm). Although this predatory fish is an ideal choice for the invertebrate aquarium, err on the side of caution and do not keep it with small fishes.*

***Right:** By the time potential predators have made up their minds which end to attack, the marine betta will have taken evasive action.*

ROYAL GRAMMA • *Gramma loreto*

This is one of the most brilliantly colored fishes to come from the Western Atlantic and Caribbean waters. The body boasts two highly contrasting colors: bright cerise on the front half and an equally bright yellow to the rear. A thin black line runs from the snout upward through the eye, and there is a black dot at the front of the long-based dorsal fin. The pectoral and pelvic fins are cerise, the caudal fin is bright yellow, and the dorsal and anal fins share both colors.

Gramma loreto shares its waters with a look-alike species, the bicolor basslet *(Liopropoma klayi)*, in which the cerise color is confined to the head region. It also lacks the eyestripe and dorsal fin dot. The royal gramma is far more frequently confused with the royal dottyback *(Pseudochromis paccagnellae)* from Indonesian and Pacific waters. Its two similar colors are separated by a hardly discernible thin white line.

Fishkeeping data

Number per aquarium: One, but established pairs could be kept in very large tanks.
Community/species tank: Community.
Swimming area: Middle and lower levels.
Food: Most foods, including live brine shrimp.
Compatibility: Peaceful but can be territorial.
Availability: Usually available (wild caught but tank-bred numbers increasing).
Captive breeding: Possible.

Below: G. loreto *dislikes any intrusion by other fishes into its chosen home. An aquarium stocked with many hideaways suits it perfectly.*

Size: *3.2 in (8 cm)*

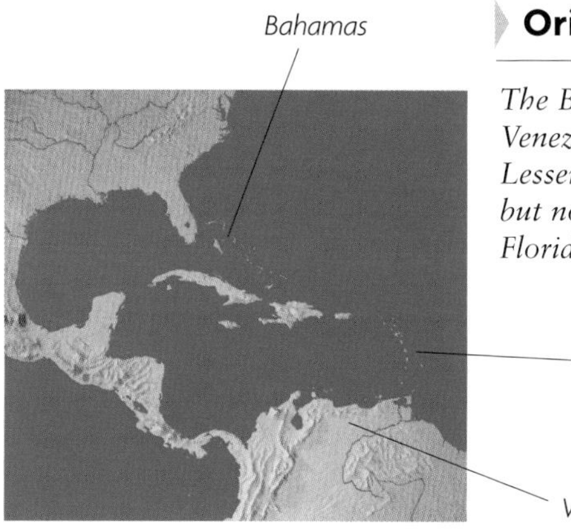

Origins

The Bahamas, Venezuela, Lesser Antilles but not Florida.

FALSE GRAMMA
PSEUDOCHROMIS PACCAGNELLAE

This brilliantly colored species, a member of the Pseudochromidae family, is almost a replica of the royal gramma (Gramma loreto)*. However, it is not quite so striking, as its dorsal and anal fins are colorless, and there is no black eyestripe. The giveaway clue to positive identification is the thin white line that runs vertically downward across the body to separate the cerise and yellow areas. It is similar in habit to the basslets, keeping close to the sea wall on reefs and swimming slowly over coral rubble in its native waters.*

Below: G. melacara *grows to 4 in (10 cm). It is a shy fish that seeks out crevices on the reef and rests upside down against the ceiling of a cave.*

Above: *The false gramma, or dottyback as it is also known, requires similar conditions to its look-alike relative, the royal gramma. It appreciates convenient retreats for security.*

SWISS GUARD BASSLET • *Liopropoma rubre*

How quick we are to assign descriptive names to fishes without giving too much thought to their suitability! This species, for example, with its alternating stripes of reddish brown and yellow arranged horizontally along the body obviously reminded its describer of the colors of the uniform of the Papal Swiss Guard in the Vatican City (even though their stripes go vertically!). There are two dorsal fins, the second of which, like the anal fin, carries a black dot. The caudal fin has black areas in its two shallow lobes, smartly terminating the striped patterning of the body. Although it is common in its native habitat, being a secretive species, it is not actually seen very often.

Fishkeeping data

Number per aquarium: Two to three.
Community/species tank: Community.
Swimming area: Middle and lower levels.
Food: All foods.
Compatibility: Generally peaceful.
Availability: Often available (wild caught).
Captive breeding: No information.

The slim body form makes "crevice creeping" an easy task.

The two dark markings on the lobes of the caudal fin are connected.

Size: *3.2 in (8 cm)*

Left: *Unsuspecting fish find it difficult to see the coral trout as it lurks under a coral overhang waiting for the next meal to come into range.*

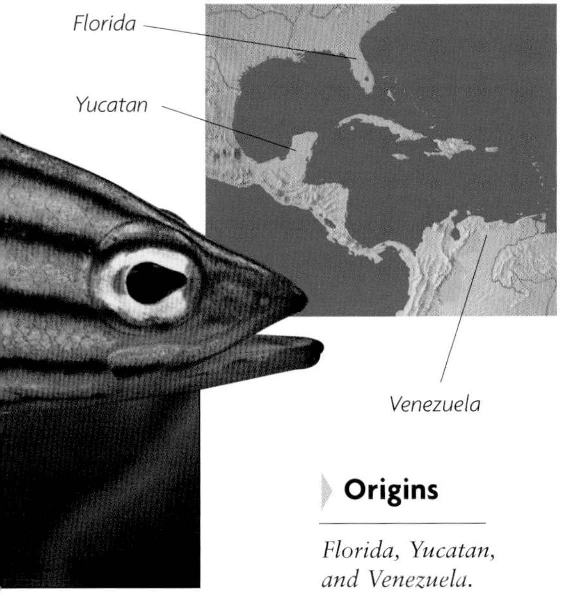

Origins

Florida, Yucatan, and Venezuela.

MINIATUS GROUPER
CEPHALOPHOLIS MINIATUS

The body and the dorsal, anal and caudal fins of C. miniatus *are bright red and covered with bright blue spots. However, the pectoral and pelvic fins are plain red. Other fishes bear a resemblance to this species, but they do not have the distinguishing rounded caudal fin. It often hides away in caves or under edges, denying the fishkeeper a view of its spectacular coloring. A large, well-filtered aquarium is essential for this fast-growing fish that can grow to 18 in (45 cm). Ranging from the Red Sea to the mid-Pacific, the predatory coral trout inhabits the coral reefs in search of a meal. Offer it smaller frozen fishes and meaty foods in the aquarium. It is safe with invertebrates, but do not keep it with small fishes.*

LYRETAIL ANTHIAS • *Pseudanthias squamipinnis*

This very large genus, containing many species, occurs in massive numbers in the wild, usually with a multitude of females acting as a harem to a very few males. If the dominant male is removed from such a group, another female will change sex to fill the gap. The Midas blenny *(Ecsenius midas)* uses its similar coloring to associate within the shoal for safety.

The body of the female fish is a golden-orange, with a pink-cerise stripe running from the eye rearward and downward across the gill cover. The fins are a golden-yellow. The male fish is more of a violet-pink with a reddish head. The gill cover decoration is yellow, and the edges of each fin are purple. The dorsal fin has an elongated third ray and the caudal fin is produced at each tip to form a lyretail.
Although obviously gregarious in the wild, within the confines of the aquarium there may not be enough room for several fish to live together peacefully, so try keeping a solitary male or just a couple of females.

Natural surfers

As these fish inhabit coral reef cliff faces near the "surf zone," the aquarium should incorporate not only suitable rocky formations, but also fairly vigorous water movement in order to make the fish feel at home. Some authorities regard anthias as "desirable but delicate" and recommend coaxing them with live adult brine shrimp to acclimatize them to captive feeding.

Fishkeeping data

Number per aquarium: One (more in a spacious species tank).
Community/species tank: Community.
Swimming area: All levels.
Food: Most foods, especially "meaty" foods and live adult brine shrimp.
Compatibility: Unless the aquarium is large enough to maintain an established "pecking-order" shoal, frequent altercations will occur between rival males.
Availability: Frequently available (wild caught).
Captive breeding: No information.

Size: 6 in (15 cm)

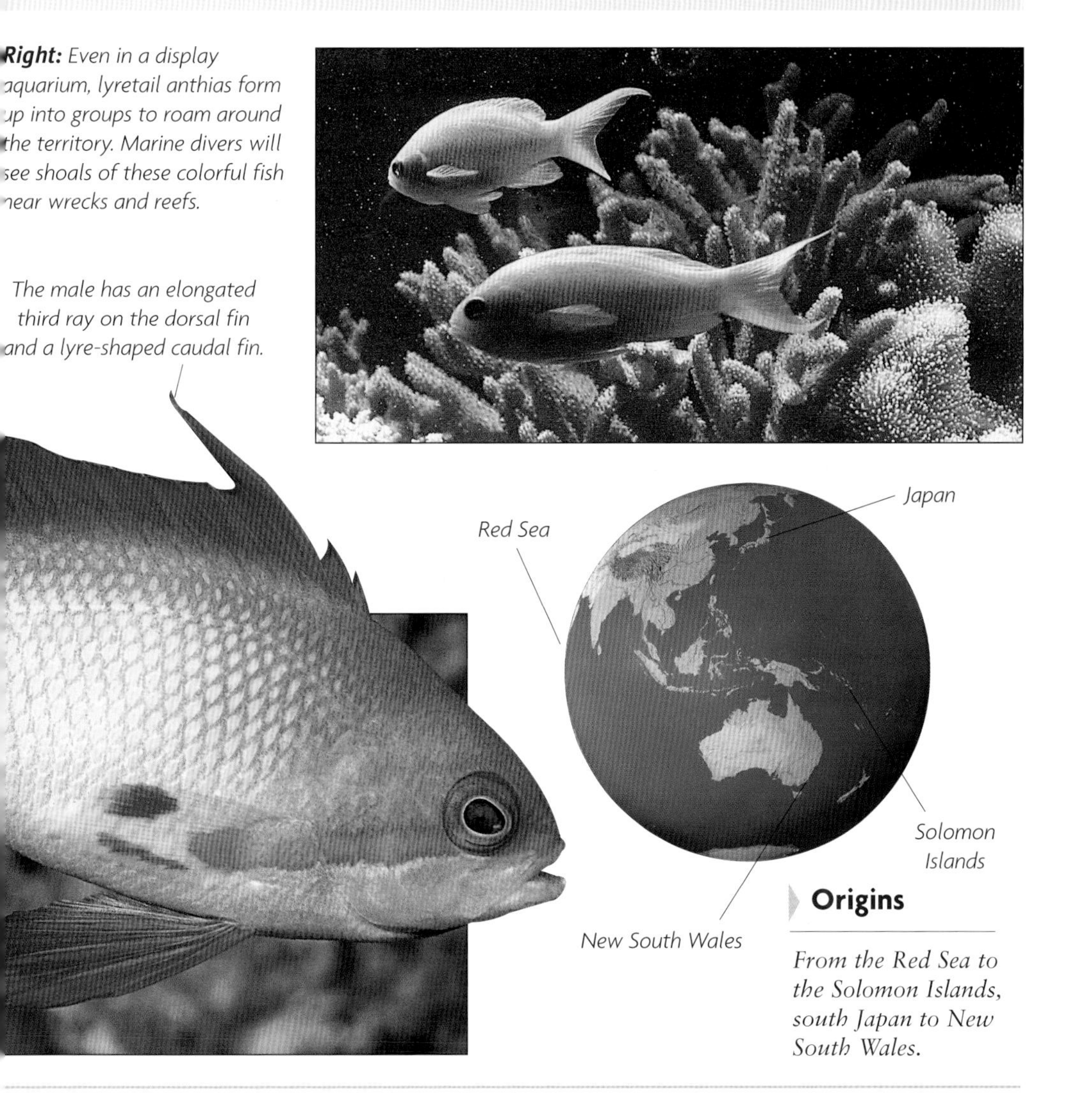

Right: *Even in a display aquarium, lyretail anthias form up into groups to roam around the territory. Marine divers will see shoals of these colorful fish near wrecks and reefs.*

The male has an elongated third ray on the dorsal fin and a lyre-shaped caudal fin.

Origins

From the Red Sea to the Solomon Islands, south Japan to New South Wales.

Blennies & gobies

Members of the Blenniidae and Gobiidae Families share many characteristics. All are small, cylindrical fish that inhabit the seabed. Generally carnivorous, they constantly dash about in search of a meal but are always ready to rush back to the sanctuary of a nearby cave or pile of rocks should they feel threatened.

Outwardly, they could be mistaken for members of the same family, but closer examination reveals quite distinct differences. Blennies generally have a continuous, long-based dorsal fin, whereas gobies have two separate dorsal fins. Another characteristic is that blennies have separate pelvic fins, while in gobies these are fused together to form a suction disc with which the fish anchor themselves into position. A further aid to positive identification is that some blennies have "cirri" – peculiar eyebrowlike growths on the head that gobies lack.

Several gobies have been bred in captivity. Many species live in substrate burrows and some even share these homes with invertebrates, such as pistol shrimps. A number of blennies are mimics; some, such as the Midas blenny *(Ecsenius midas)*, in order to associate with other fishes for safety reasons, others for more ulterior motives (especially the sabre-tooth blenny, *Aspidontus taeniatus*, which mimics the cleanerfish in order to procure a quick meal!) Always choose carefully before you buy.

The similarly shaped dartfishes have been included in this section for convenience, although they strictly belong to the Family Microdesmidae and, unlike either blennies or gobies, spend more time away from the substrate area of the aquarium. They are easily distinguished by the very long extended rays in the dorsal fin.

Above: *The decorated dartfish* (Nemateleotris *sp.*) *is a stunning sight in a marine display. The colors are intense and almost glow in the aquarium lighting.*

BICOLOR BLENNY • Ecsenius bicolor

The popular name of this species could, conceivably, be taken in two ways: either as a straightforward description of the two-part coloration of the body, or as a reference to the various color phases seen in the fish. In general, the front half of the body is dark blue and brown and the rear portion is bright yellow. The two sections of the continuous dorsal fin correspond in color to the part of the body they adjoin, while the anal and caudal fins are yellow. Two curly "eyebrows" appear in front of the eyes.

Variations in color include an all-brown form, and a dark-topped body with silvery blue lower flanks separated by a horizontal white band. The caudal fin and most rearward section of the body are yellow. Further color variations occur during spawning, when the male turns red with white transverse bars, and the female becomes yellow.

Ryukyu Islands

Maldives

Great Barrier Reef

Origins

From the Maldives, eastward and northward to the Ryukyu Islands and south to the Great Barrier Reef.

Sharp teeth

When they leave the relative safety of the substrate, this group of fishes is less likely to be preyed upon because they have canine teeth, from which they derive not only protection, but also their popular name of fang, or saber-tooth, blennies.

Size: *4 in (10 cm)*

Fishkeeping data

Number per aquarium: One – maybe more in a separate, spacious species tank.
Community/species tank: Quiet community tank or species tank with plenty of hideaways.
Swimming area: Substrate level.
Food: All foods.
Compatibility: Shy and peaceful; do not keep with larger fishes.
Availability: Generally available (wild caught).
Captive breeding: No information.

The body colors of male and female may be variable, as both change color at breeding time.

SMITH'S FANG BLENNY

MEIACANTHUS SMITHI

This very smart fish measuring 3.2 in (8 cm) has a smoky-gray-blue body topped with a white-edged, dark blue dorsal fin. The anal fin is blue. Unlike some members of the genus, the black-streaked caudal fin is rounded and not lyre shaped. There is a diagonal blue-edged black band running upward through the eye.

This genus is far more adventurous than most blennies; because it has a fully functional swimbladder, it can venture up into midwater with greater ease than its more substrate-bound relatives.

Another fang blenny (Plagiotremus phenax) *is a mimic of this species, but can be differentiated from* M. smithi *by the slightly broader dorsal fin with a more rounded front, a black-edged anal fin, and a squared-off caudal fin. It lacks the eye-bar. This differentiation is vital, because whereas* M. smithi *is a peaceful species,* P. phenax *is definitely not.*

RANDALL'S PRAWN GOBY • *Amblyeleotris randalli*

The blue-gray body is crossed vertically by seven orange-red bands. The two dorsal fins are yellow, the foremost one having a dark ocellus, or "eyespot," at its base. The pectoral, pelvic, anal and caudal fins are the same color as the body.

This body color and patterning is shared by several other species, including Steinitz' prawn goby *(A. steinitzi)*, which has a dark area above the mouth and no dorsal fin eyespot, and Magnus' prawn goby *(A. sungami)*, which has plain fins and no dark area.

Spotted prawn goby

Members of this genus live in burrows with a shrimp or prawn of the Alphaeid group. Since these invertebrates are usually blind or only partially sighted, the association works well; the prawn digs the burrow, the fish acts as lookout. The creamy-yellow body of A. guttata *(below) is covered with reddish brown spots that extend into the dorsal, anal, and caudal fins.*

Fishkeeping data

Number per aquarium: One.
Community/species tank: Community or species tank with a sufficiently deep substrate to accommodate burrowing.
Swimming area: Substrate level.
Food: Meaty foods.
Compatibility: Peaceful.
Availability: Often available (wild caught).
Captive breeding: No information.

Size: 3.5 in (9 cm)

Origins

From the Moluccas eastward to the Solomon Islands, Ryukyu Islands south to the Great Barrier Reef.

BLUE STREAK

VALENCIENNEA STRIGATA

At first glance, the delicate blue body and yellow head of this species resemble the coloration of the yellow-headed jawfish (Opisthognathus aurifrons). *However, the presence of two dorsal fins (the first often carries extended rays in adults) marks the fish as a goby, while the bright blue streak running obliquely upward across the gill cover is another distinguishing feature.*

In nature, a characteristic action of this burrowing, prawn-independent genus is its substrate-chewing, as it sifts through particles searching for tiny worms and crustaceans. It prefers a quiet community or species tank and grows to 7 in (18 cm).

V. strigata *is normally classed as a sleeper goby, a group of gobies previously referred to as Eleotridae, and usually regarded as having a sedentary lifestyle. Specimens of this genus also appear to communicate through movements of the mouth. Whether these "signals" are audible or simply visible signs is not clear.*

OLD GLORY • *Amblygobius rainfordi*

This species is very reminiscent of the Swiss Guard basslet (*Liopropoma rubre*, page 156) in its coloration, but the body ground color is a greenish yellow and the red horizontal stripes are much thinner. Two dark ocelli appear, one at the base of the second dorsal fin and another at the top tip of the caudal peduncle. The first dorsal fin and the anal, pectoral, and pelvic fins are clear. The body shape is less blunt headed than that of *L. rubre* and its symmetrical dorsal and ventral profiles indicate that it may spend as much time in midwater as on the substrate.

Brown-barred goby

The heavily built, streamlined body of A. phalaena *has a greenish gray background crossed by alternating vertical bands of brown and white. It grows to 6 in (15 cm) and lives in western Pacific waters, from Sumatra to the Great Barrier Reef and New South Wales.*

Fishkeeping data

Number per aquarium: One.
Community/species tank: Community, or species tank with a sufficiently deep substrate to accommodate burrowing.
Swimming area: Middle and lower levels.
Food: Meaty foods.
Compatibility: Peaceful.
Availability: Often available (wild caught).
Captive breeding: No information.

Size: *2.5 in (6.5 cm)*

Origins

From the Philippines southward to northwestern Australia and the Great Barrier Reef.

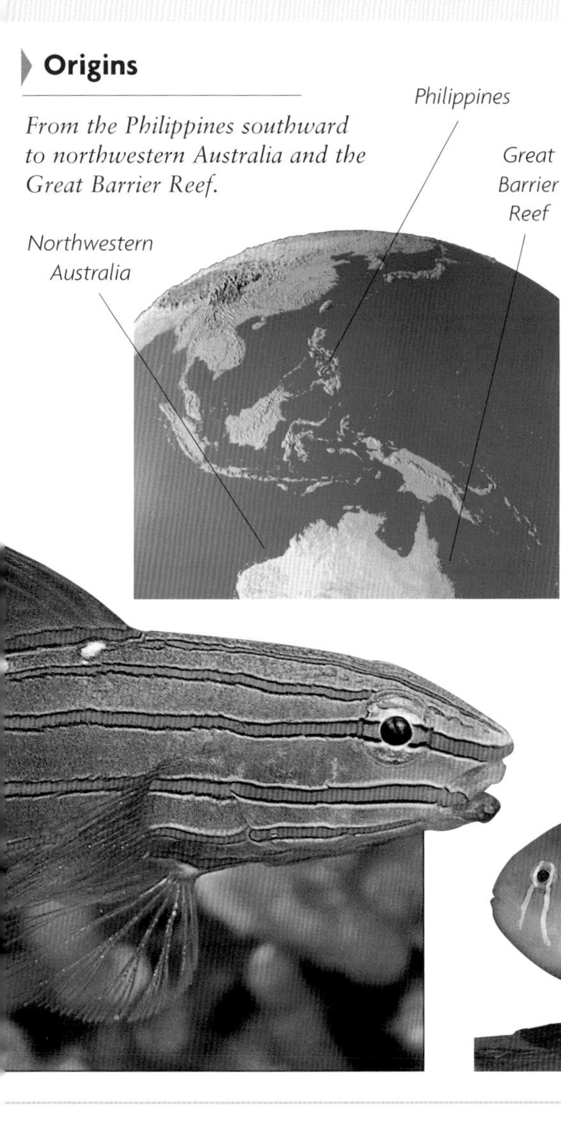

CITRON GOBY

GOBIODON CITRINUS

As its popular name implies, the body and fin color is a lemon-yellow; only the dorsal surface is a slightly darker shade. However, this plainness is alleviated by the addition of brilliant electric-blue streaks – lines along the bases of the dorsal and anal fins, two downward diagonal streaks across the rear of the gill cover and two further streaks radiating down from the eye. However, this species is variable in color and reports suggest that overall green or brown specimens occur.

This species has two distinguishing characteristics. Firstly, unlike most gobies, it is not a bottom-dwelling species, preferring to spend its time among branches of the coral heads on the reef. This trait is reflected in its alternative popular name, coral goby. Secondly, it is protected against predation by the bitter-tasting mucus layer covering its skin. This peaceful species will accept all foods and grows to 2.5 in (6.5 cm).

NEON GOBY • *Gobiosoma oceanops*

Despite the growing numbers of gobies becoming available in the hobby, the neon goby remains one of the most popular, although it could be classed as a veteran.

The dark blue body, with a white underside, is highlighted by a glowing electric-blue line running from the snout the whole length of the body. Such a simple design, yet so attractive. The species – one among a number of so-called cleaner gobies in the genus – can be distinguished from similar species by the gap on the snout between the two blue lines. In other species, the lines may be joined across the snout or have a "pip" marking in the gap.

In nature, this goby sits in crevices or caves waiting for food or, perhaps, a customer for its cleaning services. Its coloration is reminiscent of the cleaner wrasse (Labroides dimidiatus) and it has been known to "clean" divers' hands.

Unfortunately, although the neon goby makes an excellent aquarium subject, being hardy and easy to keep and breed, it is rather short lived. If you want to have them for a reasonable time, it is best to buy several juveniles rather than just one or two obviously adult fish.

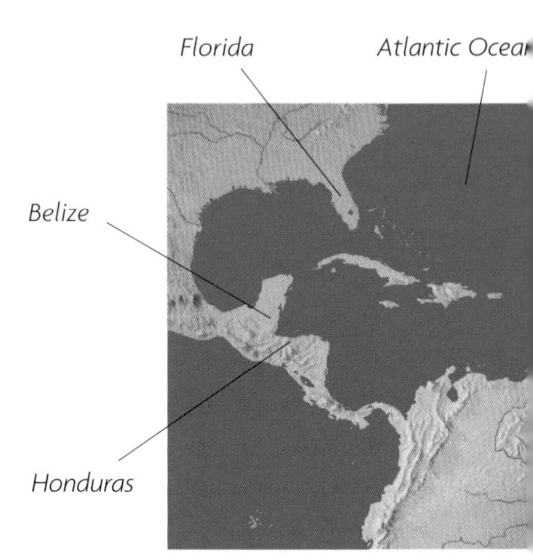

Origins

Western Atlantic waters around Florida, Belize, and Honduras.

Fishkeeping data

Number per aquarium: Two to three.
Community/species tank: Quiet community or species tank with rocky caves.
Swimming area: Substrate level.
Food: All foods.
Compatibility: Peaceful and maybe even bold with larger fish.
Availability: Generally available (tank bred).
Captive breeding: Long-standing aquarium-bred species.

Size: *2 in (5 cm)*

Breeding

One of the earliest gobies to be bred in captivity. It lays its eggs in holes in rocks, in vacant "tubes" of feather-duster worms, plastic pipes or on any convenient firm surface. The male guards the eggs. Once the fry are free swimming, they develop quite rapidly, and appear to mature within a few months.

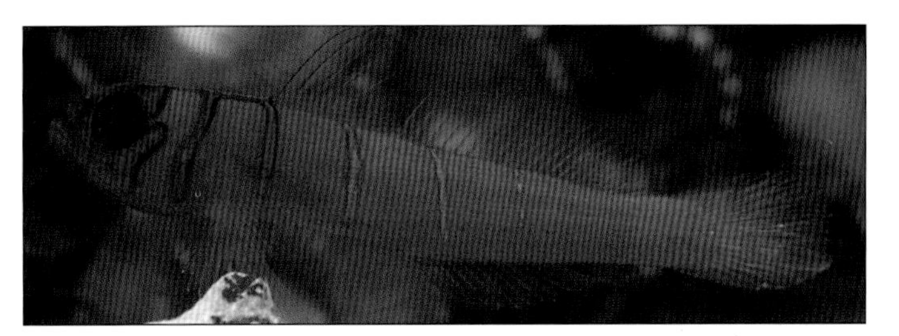

BLUE-BANDED GOBY

LYTHRYPNUS DALLI

Understandably, this gorgeously colored goby is a long-standing aquarium favorite. The body is a glowing red, overlaid with a number of vertical bright blue stripes. The first few rays of the first dorsal fin are elongated. There is a similar-looking species (L. pulchellus), *but its body is not such a deep red, and it has more vertical stripes. The blue-banded goby accepts meaty foods but prefers small live foods and grows to 2.25 in (5.7 cm). It is peaceful, although it can be territorial among its own kind.*

It might be said that this fish is the equivalent (in terms of its life span) of many of the freshwater killifishes, living, as it does, not much longer than a year or so. As its natural home is quite a distance northward from that of most "tropical" marine species, it does not require quite such a high water temperature in the aquarium, which would probably shorten its naturally brief life even more. It has been bred in the aquarium.

DECORATED FIREFISH • *Nemateleotris decora*

It is no wonder that this genus has become so popular. The fishes' colors (and their willingness to display them) are simply stunning. The creamy-yellow cylindrical body is framed by purple and red finnage, with the first rays of the front dorsal fin carried permanently erect like a banner. A purple mask covers the forehead and is led by a thin line upward over the back to join the dorsal fin.

Although happy to hover in groups in midwater, where water currents bring their food, these fish also require immediate access to safe boltholes should they feel threatened. An aquarium suitably decorated with rocks is a prerequisite to making these beautiful fishes become settled in captivity.

Fishkeeping data

Number per aquarium: Two or three.
Community/species tank: Quiet community or species tank.
Swimming area: Middle and lower levels.
Food: Most meaty foods.
Compatibility: Shy and peaceful.
Availability: Generally available (wild caught).
Captive breeding: No information.

Origins

From Mauritius eastward through the Maldives, Ryuku Islands south to New Caledonia.

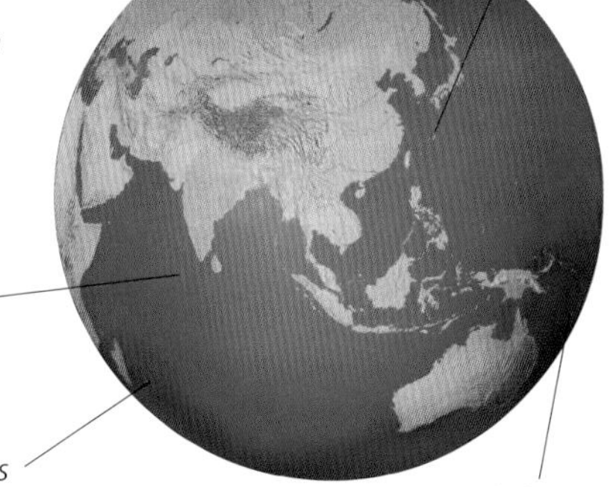

***Size:** 3.5 in (9 cm)*

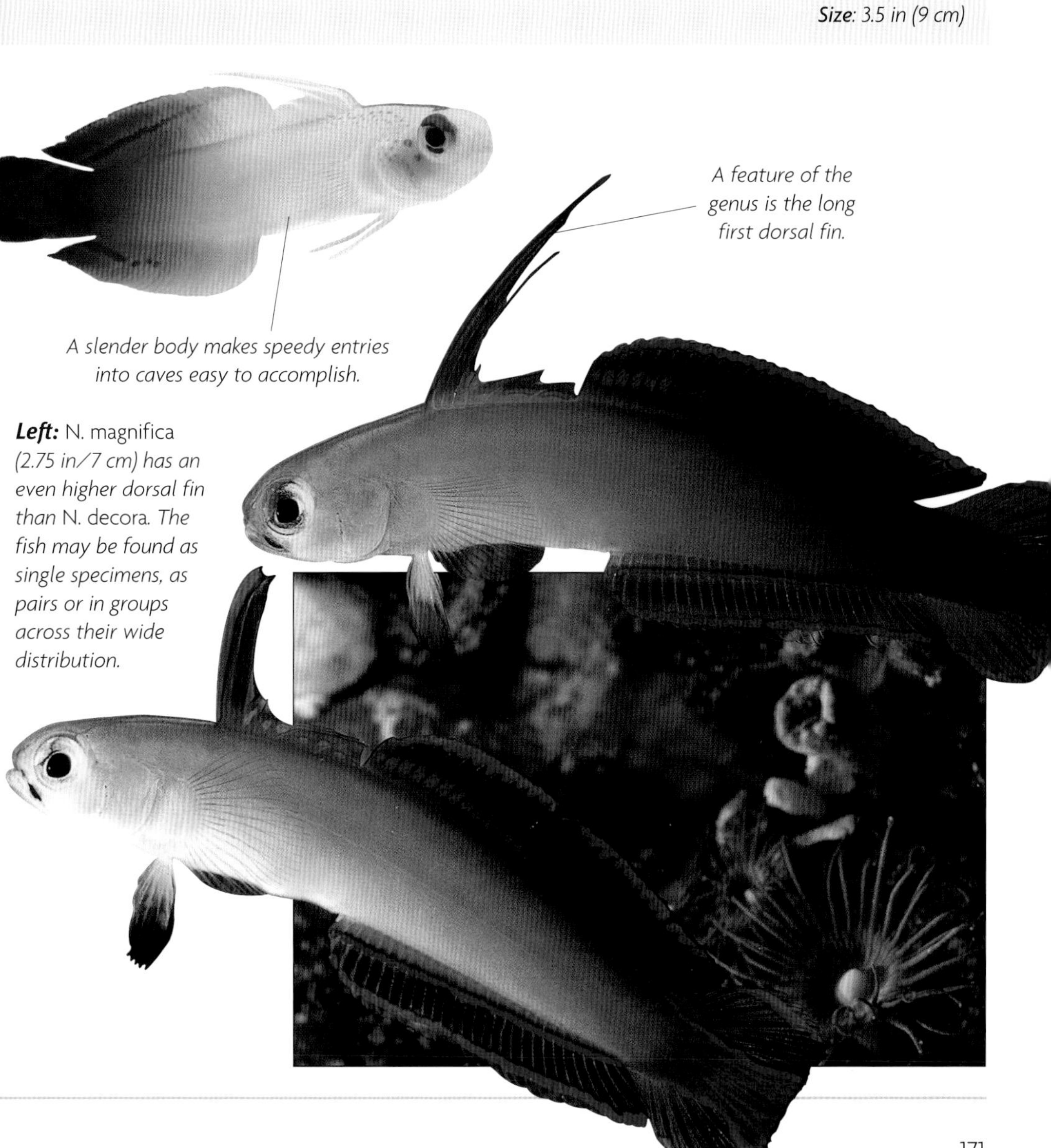

Left: N. magnifica *(2.75 in/7 cm) has an even higher dorsal fin than* N. decora. *The fish may be found as single specimens, as pairs or in groups across their wide distribution.*

Boxfishes, cowfishes, & pufferfishes

This group of fishes encompasses several families, but we have collected them together for convenience, as they share almost a common body shape and similar physical characteristics.

It is the exterior coverings of these fish that provide the interest. In the boxfishes and cowfishes there is a covering of bony plates that form an exoskeleton. The pufferfishes have two characteristic features: erectile spines and the ability to inflate their bodies to a size that will deter any foes. Some members of this group can further defend themselves when threatened or under stress by releasing self-generated toxins into the water. Handle them with care and always transport specimens alone in their own bag. Should a specimen die unnoticed in the aquarium, it is possible that the whole population of the aquarium will be wiped out as a result of the toxins from the corpse.

Despite all these warnings, do not be left with the impression that these fishes are not suitable tankmates for your aquarium. On the contrary, they soon become "tame" and get to know when it is feeding time. Due to their adult size, only juvenile specimens are likely to be available commercially, which is no bad thing for fishkeepers with modest-sized – and invertebrate-free – tanks.

Below: *Cowfishes are pretty strange-looking creatures, even more so when seen from the front! Approach the front glass of their aquarium and they will slowly come to have a look at you.*

COWFISH • *Lactoria cornuta*

Many aquarium fish are kept for their curiosity value and none is more curious than the cowfish. Here is a fish that wears its skeleton on the outside as a suit of rigid bony plates. The only apparent area for growth is the part of the fish that sticks out from the caudal peduncle. Naturally, the fish gets its popular name from the two "horns" that project from the top of a square head that rises steeply from a terminal mouth. The mouth is conveniently situated right at the bottom corner of the head, which makes for easy feeding on invertebrates on the substrate. The fish flips them up with a jet of water from its mouth and seizes them as they fall.

Two "horns" adorn the head of the aptly named cowfish.

The caudal fin (along with the dorsal fin) is the only "active" part of the fish, protruding from the bony, boxlike body.

Fishkeeping data

Number per aquarium: One.
Community/species tank: Species collection or community with very peaceful and slow-moving tankmates.
Swimming area: Middle and lower levels.
Food: Almost everything; appreciates green matter, plus shellfish. The shells help to keep the teeth worn down.
Compatibility: Do not keep with small fishes or invertebrates.
Availability: Frequently available (wild caught).
Captive breeding: No information.

Size: 18 in (45 cm)

Origins

From the Red Sea to the Marquesas and Tuamotu Islands, South Korea, South Japan, and to Lord Howe Island.

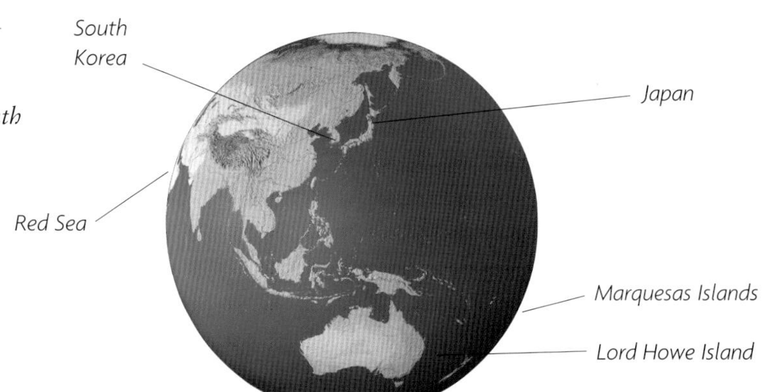

Right: *This view highlights the mouth at the front "corner" of this fascinating fish. It uses this to take invertebrates from the substrate.*

SPOTTED BOXFISH • *Ostracion meleagris*

This species was responsible for much confusion until it was realized that two differently patterned fish were not, in fact, different species, but merely male and female of the same species!

The male has the best coloration, with the two distinct areas of the body being separated by a yellow line. The dorsal surface is black, the area below the yellow line is blue and the body (excluding the blue head) is speckled with black-edged yellow spots. The pelvic fins are missing, and propulsion is achieved by the rear-located dorsal and anal fins only. The female has a black body covered with white spots.

In the wild, this species feeds on sponges and bottom-dwelling invertebrates.

The male fish has a two-patterned body, while the female (below) is uniformly spotted.

Size: 6.2 in (16 cm)

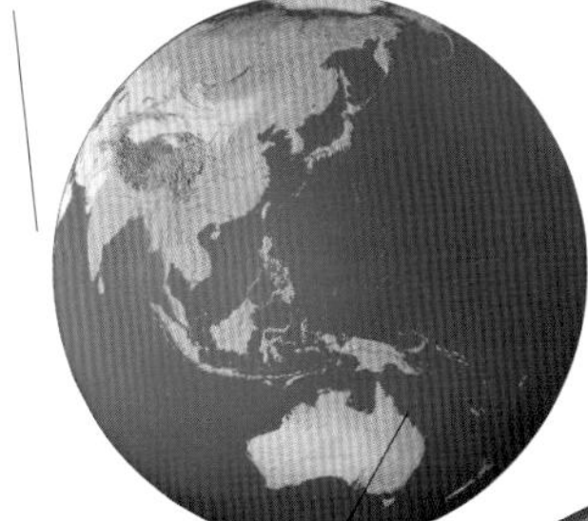

Origins

East Africa to Mexico via Hawaii, south Japan to the Great Barrier Reef and eastward to the mid-Pacific islands.

SPOTTED CUBE

OSTRACION CUBICUS

Young specimens of these fish from the Red Sea and tropical Indo-Pacific, resemble little yellow dice. Their bold black spots on a bright yellow background may serve as a warning to predators. When attacked or frightened, the fish will release poisonous ostracitoxin into the water. Adult fish, measuring 18 in (45 cm) lose much of their bright livery, becoming a drab brown. This fish is relatively easy to keep and will enjoy most foods that are offered. However, take care when introducing it into the aquarium so that it does not become unduly stressed. If this happens it will secrete its poison into the water, thereby killing all the tank inhabitants, including itself. For the same reason, it should not be housed with aggressive or boisterous fishes.

Fishkeeping data

Number per aquarium: One.
Community/species tank: Species collection, or community with very peaceful and slow-moving tankmates.
Swimming area: Middle and lower levels.
Food: Almost everything; appreciates green matter, plus shellfish. The shells help to keep the teeth worn down.
Compatibility: Do not keep with small fishes or invertebrates.
Availability: Frequently available (wild caught).
Captive breeding: No information.

SPINY BOXFISH • Chilomycterus schoepfi

The body shape is much like a horizontally elongated teardrop, with a slightly flattened, broad top and the large eyes set high up. The yellow body is covered with a series of dark brown stripes, and several large, dark ocelli are distributed over the body. The spines are held erect as a permanent deterrent to would-be predators and this stands the fish in good stead, as it does not inflate itself as do most members of this group. It is not regularly seen at aquatic outlets.

Origins

Common in Florida (and occasionally northward), plus coastal waters south to Brazil (excluding the Caribbean islands).

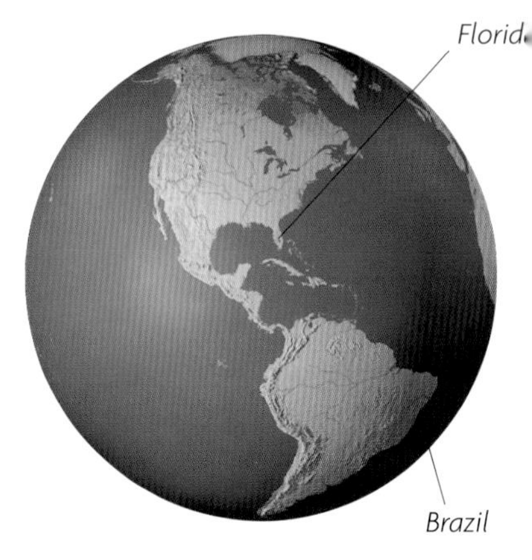

The relatively shallow body allows the fish to lurk in caves and narrow crevices.

Size: 20 in (50 cm)

Above: *These smoothed-down spines can soon be erected to form a spiky defence.*

Fishkeeping data

Number per aquarium: One.
Community/species tank: Species collection, or community with very peaceful and slow-moving tankmates.
Swimming area: Middle and lower levels.
Food: Almost everything; appreciates green matter, plus shellfish. The shells help to keep the teeth worn down.
Compatibility: Not with small fishes or invertebrates.
Availability: Frequently available (wild caught).
Captive breeding: No information.

BALLOONFISH

DIODON HOLOCANTHUS

The body has a yellowish tinge with a white ventral surface. A number of short dark bars straddle the dorsal surface, while a much longer one covers the eye and gill cover. The dorsal, anal and caudal fins are yellow. A feature of the Diodontidae family is the fusing together of the two front teeth on each jaw ("diodon," meaning two teeth) to form a bony beak for crushing shelled invertebrates. The fish is a poor swimmer and its large eyes may denote that it does its feeding more by night than by day.

One problem with all "inflatable" fish is that they can inflate with air, rather than water, especially if lifted from the water. In such cases, the fish may find it very difficult to deflate itself afterward.

Balloonfish grow to 20 in (50 cm) and are found worldwide in all warm seas. Keep them in a species collection or a community with very slow-moving tankmates but not with smaller fishes or invertebrates. Provide the same diet as for the spiny boxfish.

Handle with care

Stress will cause all the porcupinefishes and burrfishes to exude toxins and thus "self-destruct." Do not "net" them (watch out for those spines!) or expose them to the air – they might "inflate." Capture them in a water-filled plastic bag.

BLACK-SADDLED TOBY • *Canthigaster valentini*

The white body of this fish has unusual coloration imposed upon it. Four dark, triangular blotches appear on the dorsal surface, from the top of the head to the caudal peduncle, the middle two extending down the flanks almost to the ventral surface. Additionally, the lower half of the body is peppered with yellowish brown spots, and the yellow caudal fin has a black top and bottom edge.

Distinctive as it may sound, this patterning is not exclusive to this species. The black saddle filefish *(Paraluteres prionurus)* has exactly the same markings, with the exception of the caudal fin, which lacks the black top and bottom edges.

Males surround themselves with a harem of females with whom they mate on a rotational and regular basis; eggs are laid in a nest of algae.

The dark markings reach further down the body than on the similar-looking species, Canthigaster coronata.

Fishkeeping data

Number per aquarium: One.
Community/species tank: Species collection, or community with very peaceful and slow-moving tankmates.
Swimming area: Middle and lower levels.
Food: Almost everything; appreciates green matter, plus shellfish. The shells help to keep the teeth worn down.
Compatibility: Not with small fishes or invertebrates.
Availability: Frequently available (wild caught).
Captive breeding: No information.

***Size:** 4 in (10 cm)*

Wide distribution including the Red Sea to Tuamotu Islands, South Japan to Lord Howe Island.

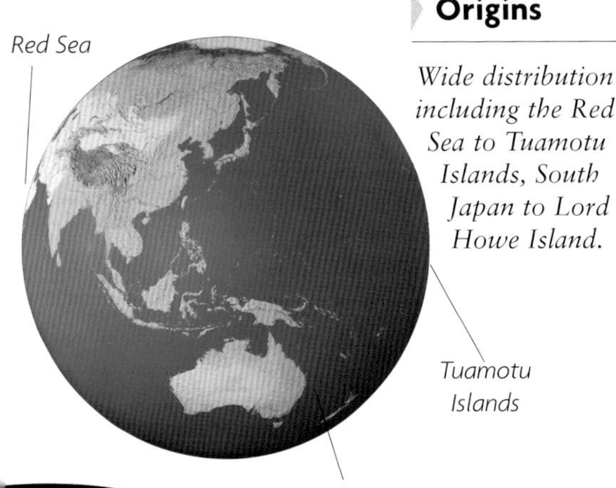

SHARPNOSED PUFFER

CANTHIGASTER SOLANDRI

This spectacularly patterned fish from the Indo-Pacific and Red Sea has a gold-brown body and a caudal fin covered with pale spots. A blue wavy line replaces the spots on the upper part of the body and a large white-edged black spot appears at the base of the dorsal fin. The fish swims with its caudal fin folded. The pelvic fins are absent. In the aquarium, C. solandri *is peaceful, except toward members of its own kind, but do not keep it with invertebrates, as it will eat them. Feed it on finely chopped meat foods. It grows to 2 in (5 cm).*

***Above:** Do not give pufferfish the opportunity to live up to their name. Avoid exposing them to the air, and when handling them, transfer them in water-filled plastic bags and not via "open-air" nets.*

Miscellaneous species

This final section of the book features fish from a number of families. Cardinalfishes (Apogonidae) are ideal for the marine aquarium. They are undemanding, and amenable to aquarium conditions and foods. A common characteristic is the presence of two separate dorsal fins. Hawkfishes (Cirrhitidae) like to perch on a rocky outcrop as they await any passing prey, typical behavior of species lacking a swimbladder. Hawkfishes should settle down quickly in captivity, but mixing hawkfish species in the tank may lead to quarrels.

The slow-moving lionfishes (Scorpaenidae) present a false picture of serenity. A sudden gulp from the huge mouth can engulf any passing prey and a sting from their venomous spines can seriously incapacitate the careless handler. As these fish are naturally active at dusk and dawn, the aquarium need not be too brightly lit. Filefishes (Monocanthidae) have two dorsal fins, the posterior one being used with the anal fin for propulsion. The skin is rough to the touch and often causes problems when netting; using a plastic bag to capture them is recommended.

Jawfishes (Opisthognathidae) need a reasonably "quiet" aquarium with peaceful tankmates, and a soft substrate in which they can build a burrow. They normally hover vertically, tail-down, around the burrow entrance into which they can retreat – backward – like lightning. Dragonets (Callionymidae) are hard to see against the substrate, where they make their permanent living space. They need copious amounts of live foods; do not keep them with sizeable tankmates who would out-compete them in the hunt for food.

Above: *With its venom-filled spines, the lionfish* (Pterois volitans) *is a favorite public aquarium exhibit, where its graceful drifting form of locomotion belies a more sinister motive – looking for food.*

BANGGAI CARDINALFISH • Pterapogon kauderni

Every so often, a species comes along that finds instant appeal among fishkeepers, and so it is with the Banggai cardinalfish. It has striking silver-and-black coloration, very reminiscent of the freshwater angelfish, *Pterophyllum scalare,* with one or two extra adornments. The top and bottom parts of the caudal peduncle have a white-edged black stripe that continues to the tips of the caudal fin. The large pelvic fins, together with the anal fin, seem to mirror the two separate dorsal fins, while the whole body is overlaid with white dots.

Discovered in 1990 in the Banggai Islands of Sulawesi in Indonesia, this fish has proved to be extremely amenable to life in captivity and many public aquariums are able to breed them freely (they are mouthbrooders). This means that future supplies to the aquatic trade should be secure, thus allowing the original wild stock to continue their natural existence without the threat of fish collectors hunting them down. In the light of this, do not be surprised if home aquarium breeding suddenly takes off.

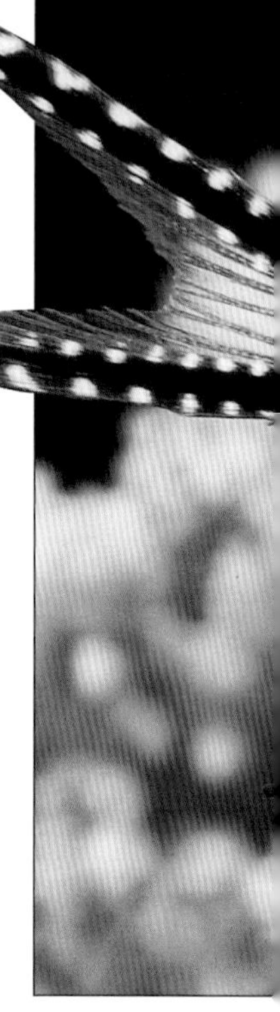

Fishkeeping data

Number per aquarium: Two or three.
Community/species tank: Species or community but not with large fishes.
Swimming area: Middle and lower levels.
Food: All "meaty" foods, fresh or frozen, plus live brine shrimp.
Compatibility: Peaceful, sometimes a little quarrelsome between themselves.
Availability: Recent introduction but plentiful (wild caught but soon tank bred).
Captive breeding: Regularly in public aquariums and increasingly in hobbyists' tanks, too.

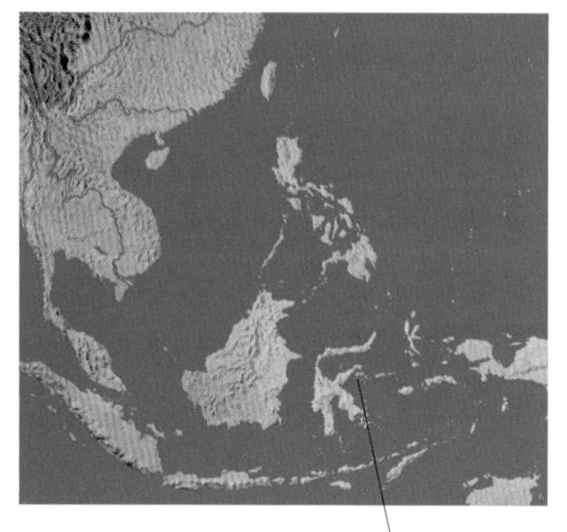

Origins

The Banggai Islands off Sulawesi in Indonesia.

Size: 3.2 in (8 cm)

White spots are part of the natural coloration of this fish – not symptoms of disease!

Above: A square, deeper jaw usually indicates a male incubating a new batch of eggs in his throat.

Breeding Banggai cardinalfishes

It is sometimes possible to acquire proven pairs, but it is best to obtain a small group and hope that there is a male and female among them and that pairing occurs.

*Eggs are produced and fertilized in the normal way for egglayers, but are then taken into the mouth of the male for incubation. Even after hatching, the juvenile fishes remain in the mouth for protection from predators. When they emerge – as tiny, fully formed replicas of the adults – juveniles immediately seek the protection of the long-spined sea urchins (*Diadema sp.*) with which they associate in the wild. It is not essential to keep sea urchins in order to rear this species successfully, but if they are not available, separate the juveniles from the adults, at least until they have grown larger. Some aquarists use artificial urchins with good results. Juvenile Banggai cardinals will sometimes protect themselves in anemones if urchins are not present in the aquarium.*

Newly hatched brine shrimp and very finely chopped mysis shrimps, etc., are good foods for these relatively large juveniles. As they grow, offer them larger food particles.

Breeding this, and similar, species can be carried out in an aquarium as small as 20 gallons (90 liters). This is due to the very small brood sizes and the fact that you can easily separate adults from juveniles by using a partition that allows a flow-through of aquarium water.

PYJAMA CARDINALFISH • Sphaeramia nematoptera

In common with the previously described species, the body shape of this fish is quite stocky and deep, with the dorsal and ventral contours being quite convex. The patterning is quite unusual in that the dark-spotted rear half of the body is separated from the unadorned front section by a broad, dark brown vertical band. The front dorsal fin has some patterning and the large pelvic fins are yellowish edged with white. The large eye, indicative of a nocturnally active species, is red-rimmed with two vertical white stripes.

Origins

From Java to Papua New Guinea, and Ryukyu Islands south to Great Barrier Reef.

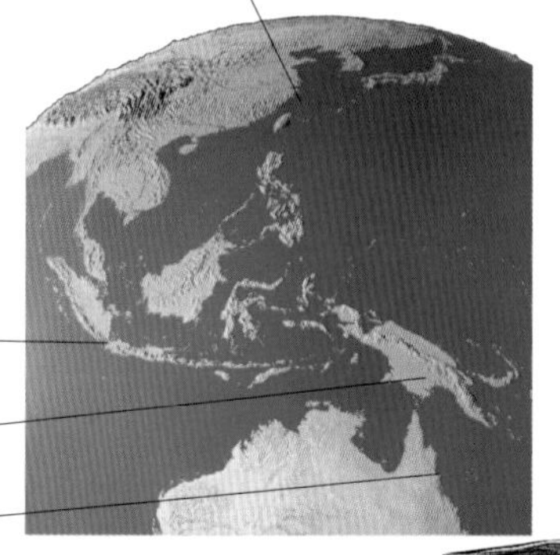

Fishkeeping data

Number per aquarium: Two or three.
Community/species tank: Community, but not with large fishes.
Swimming area: All levels.
Food: All "meaty" foods, fresh or frozen, plus live brine shrimp.
Compatibility: Peaceful, shy, and sometimes a little quarrelsome between themselves.
Availability: Often available (wild caught).
Captive breeding: No information but may occur unnoticed.

Size: 3.2 in (8 cm)

The pyjama cardinalfish's body is bizarrely marked with different zones of patterning.

YELLOW-STRIPED CARDINALFISH

APOGON CYANOSOMA

True to its common name, the streamlined body carries five or six horizontal yellow lines along the flanks. The eye is large. When the fish is viewed from the side, the pectoral, pelvic, and the first (of two) dorsal fins all line up vertically. All the fins are colorless. A similar-looking species, the blue-spot cardinalfish (A. nitidus)*, is found over much of the same distribution area (Red Sea to the Marshall Islands and south Japan to the Great Barrier Reef), but it has a somewhat brown tinge to the yellow areas and a dark line running centrally through the caudal fin.*

The yellow-striped cardinalfish is peaceful, sometimes shy and will not bother other fishes. Keep it in a community tank but not with large fishes. It accepts all "meaty" foods, fresh or frozen, plus live brine shrimp, and grows to 3.2 in (8 cm).

LONGNOSED HAWKFISH • *Oxycirrhites typus*

At one time, this was the only hawkfish likely to be found in dealers' tanks. The long snout and jawline, together with the squares of bright red covering the whitish body, make the longnosed hawkfish instantly recognizable. The long-based dorsal fin has a spiny front section, and tiny growths at the tips of the dorsal spines are often visible. These growths, known as cirri, account for the derivation of the family name, Cirrhitidae.

This species, along with others in the genus, is not an active fish, spending its time on an outcrop of coral or any suitable aquarium decoration, waiting for a feeding opportunity to present itself. This apparently peaceful occupation may seem innocent enough, but small fishes and non-sedentary invertebrates may be at risk.

Fishkeeping data

Number per aquarium: One.
Community/species tank: Community but not with very small fishes or invertebrates.
Swimming area: Middle and lower levels.
Food: All foods.
Compatibility: Fairly peaceful, may be territorial; certainly an opportunistic, rather than predatory, feeder.
Availability: Regularly available (wild caught).
Captive breeding: No information.

Origins

The natural distribution of the longnosed hawkfish ranges from the Red Sea to Panama, including Japan, Hawaii, and New Caledonia.

Hawaii
Panama
Red Sea
Japan
New Caledonia

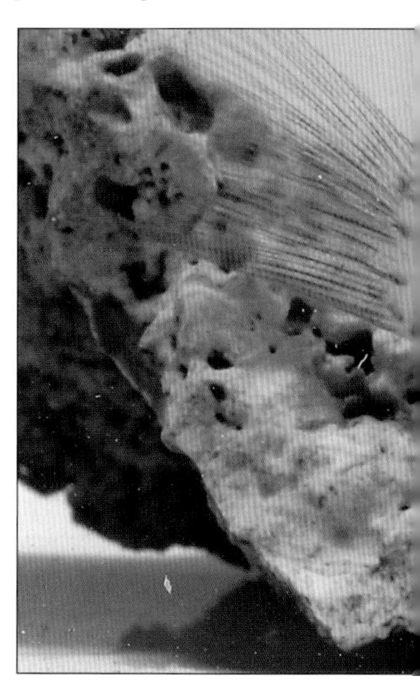

Size: *5 in (13 cm)*

Left: *The flat lower contour of the body makes perching on rocky outcrops easy, as the longnosed hawkfish waits for a passing meal.*

Below: *A red line on the lower edge of the jaw may indicate a female.*

ARC-EYE HAWKFISH • *Paracirrhites arcatus*

With its high, arched back and flatter ventral surface, the body contour of this fish is not entirely symmetrical, but obviously an efficient design for a fish that does not do a lot of swimming around. The head and dorsal surface are red, the flanks below the median line are yellow and separated by a bright white band (not present in all cases) running from mid dorsal fin to the end of the caudal peduncle. Some radiating, red-edged, blue bands emerge from the rear edge of the gill cover. All the fins are yellow. The species takes its common name from a blue, red, and yellow-edged dark area on the head just above, and to the rear of, the eye. Other species share this decoration.

The "eye patch" is not exclusive to this species, but the white body stripe (when present) helps with positive identification.

On the lookout

The high-set eyes of the hawkfishes enable them to spot potential prey easily. Their body shape is ideally suited to their sedentary life; they prefer their meals to come to them, rather than chase their food.

Fishkeeping data

Number per aquarium: One.
Community/species tank: Community but not with very small fishes or invertebrates.
Swimming area: Middle and lower levels.
Food: All foods.
Compatibility: Peaceful.
Availability: Often available (wild caught).
Captive breeding: No information.

Size: *5.5 in (14 cm)*

Origins

From East Africa to Hawaii, south Japan to Norfolk Island.

SCARLET HAWKFISH

NEOCIRRHITES ARMATUS

This extremely attractive 3 in (7.5 cm)-fish from the Central and Western Pacific is much sought after by aquarists and usually commands quite a high price as a consequence. It is not a good swimmer, and in the wild it spends much of its time resting on rocks and gorgonians waiting to ambush small crustaceans and plankton. Fortunately, it is not a threat in the home aquarium and seems to fare best in a mixed fish and invertebrate system, where it also tends to look more natural. It accepts most marine frozen and live foods of a suitable size. Once settled in the aquarium it will also accept marine flake. An ideal community addition.

Above: *"Hawkfish" is an apt name for these species, which are always perched on rocks on the lookout for food.*

DWARF LIONFISH • *Dendrochirus zebra*

Such are the proportions of the intricately patterned fins, that it is often hard to determine the exact body shape of these fishes. In fact, the body is sturdily built, with a high dorsal arch, and, unsurprisingly for a genus that spends some time around the substrate level of the aquarium, a flattened ventral surface.

Deep reddish brown bands, alternating with lighter brown, cross the body vertically, and this patterning appears to extend concentrically into all fins. The large, fan-shaped pectoral fins have connecting tissue membrane right to the tips of the fin rays. The first few rays of the dorsal fin are separate, and there is a pair of "hornlike" growths above the eyes. A distinguishing characteristic of this species is the dark ocellus spot on the lower part of the gill cover.

Japan

Samoa

Lord Howe Island

Origins

From South Africa to Samoa, south Japan to Lord Howe island.

Below: *The pectoral fins have much more connecting tissue between the spines than is seen in other species in this family group.*

The coloration of the winglike fins is just as impressive as that of a butterfly.

Size: 7 in (18 cm)

Fishkeeping data

Number per aquarium: One.
Community/species tank: Species.
Swimming area: Middle and lower levels.
Food: A wide appetite for all foods, especially "meaty" derivatives.
Compatibility: Predatory with small fishes and some invertebrates.
Availability: Often available (wild caught).
Captive breeding: No information.

OCELLATED LIONFISH

DENDROCHIRUS BIOCELLATUS

Compared to related Pterois *species, the smaller (around 4.7 in/12 cm)* Dendrochirus *is poorly equipped when it comes to extended finnage. However, the ocellated lionfish has the distinction of being the only lionfish with "eye-spots" on the rear soft portion of the dorsal fin, and the large butterfly-wing pectoral fins have dark concentric rings. Another exclusive feature is the presence of two barbels on the upper jaw.*

The fish is not often seen in the general day-to-day activity on the reef due to its habit of lurking (often upside down) beneath overhangs in the coral, only venturing out to feed at night. It may lose out in the feeding race if kept with more lively fish, and shrimps, etc., in a reef tank may be at risk.

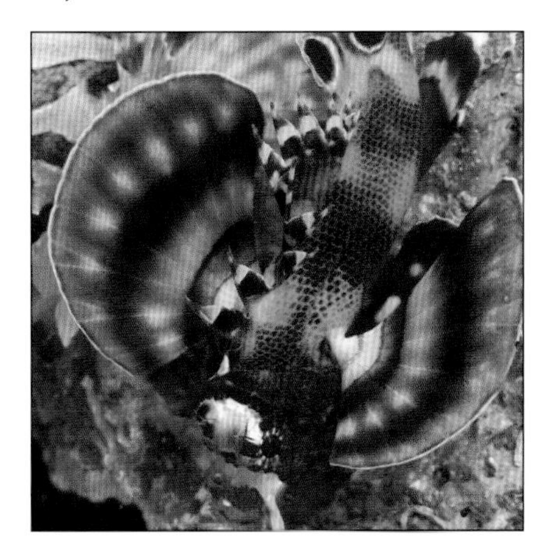

LIONFISH • *Pterois volitans*

Unlike the previous species, which is classed as a dwarf lionfish by fishkeepers, this is the real thing and quite fantastic in all its enlarged beauty.

Once again, the reddish brown-striped body is surrounded by a huge "plumage" of fins. Although separated, each one of the first dozen rays of the dorsal fin has a backing of membranous tissue, giving them the appearance of feathers. Similarly, the rays of the huge pectoral fins are separate, with decorated tissue along their length. The function of these large fins is to herd prey into a convenient area for feeding on, and it is not hard to imagine a number of lionfish acting collectively, by using their impressive fins as capes to round up their next meal.

This species, usually seen "drifting with intent" around rock faces and under coral ledges in public aquariums, seems to spend more time off the seabed than its smaller relatives.

Below: *The large mouth is just waiting to engulf any passing prey – in one gulp!*

Origins

From Malaysia to Pitcairn Island, south Japan to Lord Howe Island.

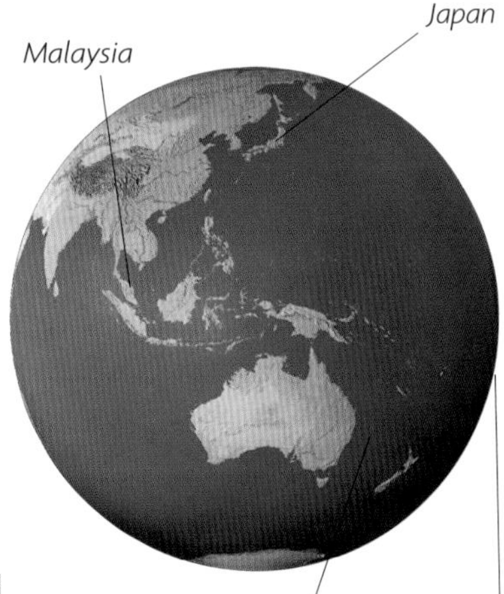

Fishkeeping data

Number per aquarium: One.
Community/species tank: Community (with large fish) or species tank.
Swimming area: All levels.
Food: All "meaty" foods.
Compatibility: Predatory.
Availability: Regularly available (wild caught).
Captive breeding: No information.

Size: *15 in (38 cm)*

Keep your hands well away from the venom-filled fin rays.

TASSELLED FILEFISH • *Chaetoderma pencilligera*

It is hard to describe the characteristics of a species that frequents floating seaweeds and models itself accordingly. The body shape, once determined, is greyish white with a few brown blotches on the lower flanks and a couple of dark spots higher up. Its outline and actual surface is further disguised by a covering of tassel-like appendages, and a number of thin brown lines also cross the body in a branchlike network. The fins are mostly clear with a little speckling, and the caudal fin is attached to the body by a very short and narrow caudal peduncle. The pelvic fins are but a single spine. The tiny snout is beaklike, ideal for browsing and grazing on algae and polyp growths.

Once acclimatized to aquarium foods, this fish is a popular favorite. Fortunately, it is unlikely to attain its full natural size in the home aquarium.

The tassel-like appendages disguise the fish's presence among seaweeds.

Fishkeeping data

Number per aquarium: One.
Community/species tank: Community.
Swimming area: Middle and lower levels.
Food: Most foods (including vegetable matter and meaty foods) once acclimatized. If necessary, offer live foods at first.
Compatibility: Fairly peaceful, but may nibble enquiringly at other fish without causing physical damage – only annoyance.
Availability: Often available (wild caught).
Captive breeding: No information.

***Size:** 10 in (25 cm)*

Japan

Malaysia

Great Barrier Reef

Origins

Waters around Malaysia, south Japan and the Great Barrier Reef.

PINK-TAILED FILEFISH

PERVAGOR MELANOCEPHALUS

The tail of this fish is pink to orange-red, and the body shades from orange-green to a deep purple. The dorsal fin may be limited to just one or two strong rays. A dark bar extends upward from the pectoral axil to a point behind the eye. This slow-moving fish from the tropical Indo-Pacific is quite secretive by nature. Its main habitat is among weed beds and algal growths, where its body color gives it an ideal camouflage.

There may be some initial difficulties when the fish is first introduced into the aquarium, but then it usually settles in well and will accept all the usual aquarium foods. Like most filefishes, it grows quite fast, but not too large for the average aquarium, reaching 6 in (15 cm). Provide plenty of rocks and hiding places into which the fish can retreat at night.

YELLOW-FACED JAWFISH • *Opisthognathus aurifrons*

For a good look at this species you will need to be patient until it is confident enough to emerge fully from its burrow. Then you can see that the long, cylindrical body is a most delicate blue, with a bright yellow, blunt head. The dorsal fin runs almost the complete length of the body and is mirrored by the anal fin, which occupies two thirds of the ventral length. The pectoral fins are yellow, the pelvic and caudal fins blue. The dark eye is very large and the mouth, terminally situated at the tip of the blunt snout, is ideally placed for shifting grains of substrate material during excavation of the burrow.

Sex differentiation is difficult, but should breeding occur, then it is reported that the male takes on the incubating duties of this mouthbrooding species.

Origins

Opisthognathus *contains several species other than this one and all are found in waters ranging from Florida and the Bahamas to Barbados and Venezuela.*

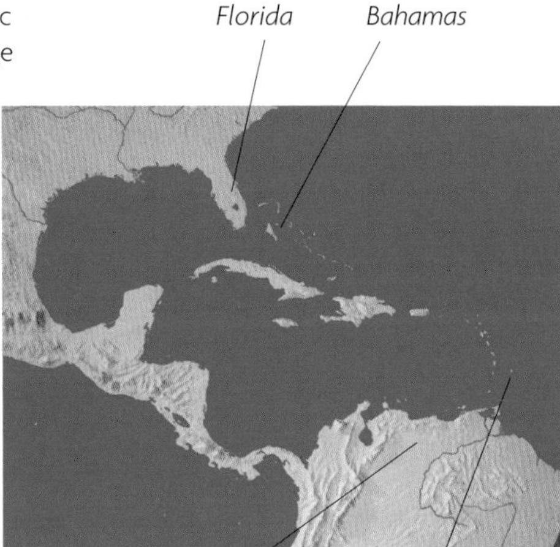

When it ventures outside its burrow, the yellow-faced jawfish "hovers" just above its safe retreat.

Fishkeeping data

Number per aquarium: One (possibly more in a species tank).
Community/species tank: Community or species tank.
Swimming area: Lower levels.
Food: Meaty foods. It is a good idea to mince these up and distribute them near to the fish's burrow.
Compatibility: Peaceful and shy.
Availability: Regularly available (wild caught).
Captive breeding: No information.

Size: *4 in (10 cm)*

The rear blue portion of the fish's body is usually hidden from view as it sits vertically in its burrow.

Above: *Keeping the burrow entrance free from falling particles is a constant chore, but having a muscular mouth helps the yellow-faced jawfish to cope with the problem.*

MANDARINFISH • *Synchiropus splendidus*

The scaleless body of this fish is covered with an amazing tangle of contrasting colors that obviously reminded the original describer of the brilliantly colored robes worn by ancient Chinese dignitaries. Golds, reds, and greens are all intertwined across the body; nor are the fins excluded from this rainbow effect, having purple edges in addition to those colors spreading from the body. The caudal fin has gold lines radiating from the caudal peduncle, and the large pelvic fins are decorated with purple/green spots. The first ray of the first dorsal fin is projected and carried like a flag.

Origins

Waters around the Philippines and Java, from the Ryukyu Islands south of Japan to the Great Barrier Reef.

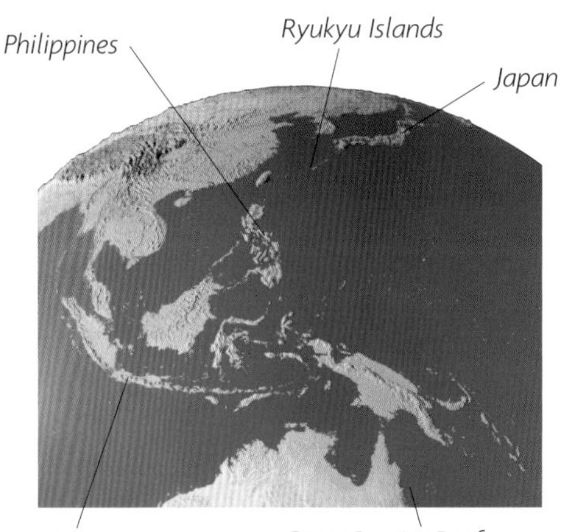

***Size:** 2.4 in (6 cm)*

Fishkeeping data

Number per aquarium: One or two.
Community/species tank: Suitably stocked community aquarium, but a species tank is probably better.
Swimming area: Lower level.
Food: Requires copious amounts of live foods.
Compatibility: Shy, but constantly on the lookout for food.
Availability: Often available (wild caught).
Captive breeding: Has spawned in captivity but not a regular occurrence, and no reports of surviving fry are known.

Breeding

From very limited reports of aquarium breeding, it seems that internal fertilization of eggs occurs before their release as floating eggs into open water. While this may be true, there are no reports of any viable fry being raised to maturity.

PICTURESQUE DRAGONET

SYNCHIROPUS PICTURATUS

Being blunt-headed, flat-bottomed, and almost cylindrical, the body shape of this fish is very similar to that of the better-known bottom-dwelling species, such as blennies and gobies. There are two dorsal fins, the first of which is usually carried like a banner.

The basic body color is light green, but this is overlaid with patches of dark green ringed with gold, black, and blue. These blotches also appear in the dorsal, anal, and pelvic fins, while the caudal and pectoral fins are a plain yellowish green. The high-set eye has a gold and black rim.

Males are generally equipped with brighter colors and extensions to the dorsal and anal fins. There is some belief that any highly colored species (fish or invertebrates) are necessarily poisonous and advertise this fact accordingly; whether this toxicity risk actually applies to the mucus of this species is subject to conjecture. S. picturatus *requires the same general care in the aquarium as the mandarinfish and also grows to 2.4 in (6 cm).*

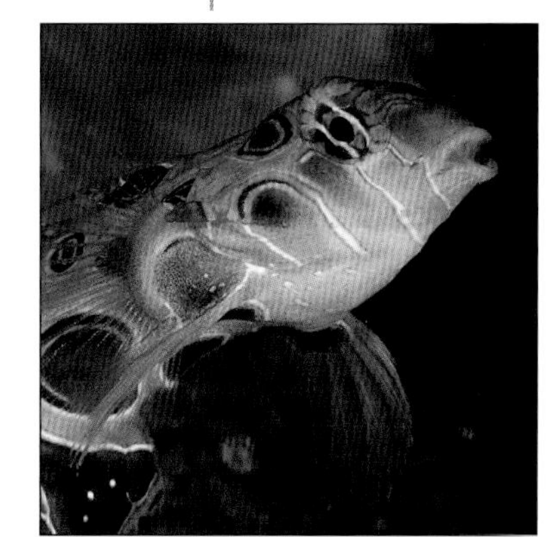

GENERAL INDEX

Page numbers in **bold** indicate major entries; *italics* refer to captions, annotations, and panels; plain type indicates other text entries.

A

B

C

D

E

F

FISH INDEX

A

O

P

R

S

T

U

V

W

Z

PHOTO CREDITS

The publishers would like to thank the following photographers for providing images, credited here by page number and position: (B) Bottom, (T) Top, (C) Center, (BL) Bottom left, etc.

Aqua Press (M-P & C Piednoir): 8, 32-33(C), 39(BL), 93(R), 99(R), 101(BR), 107, 109(TR), 111, 113(BR), 115, 123, 127, 129, 131(T), 137, 139(T), 141(B), 149(BR), 153(BR), 157(T), 181, 193(B), 197(B), 199

Bioquatic Photo – Alf J. Nilsen (bioquatic@biophoto.net): 71

Bruce Coleman Collection: 200-201 (Jane Burton)

Les Holliday: 85(T), 177(BR)

Photomax (Max Gibbs): 90-91, 92, 94-95, 96-97, 98-99, 100-101, 102(L), 102-103(C), 106(T,B), 108-109, 110-111, 112, 112-113, 114, 118, 119(BL), 120-121, 122, 125, 126-127, 128-129, 130-131(B), 132-133(B), 133(T), 136-137, 138-139, 140-141(T), 142, 143, 146-147(T), 147(B), 148-149, 152-153, 154, 154-155(C), 155, 156-157(B), 158-159, 162-163, 163(TR), 164(T), 164-165, 165(BR), 166(T), 166-167, 167(BR), 168-169, 169(TR), 170, 171(BR), 174, 176, 178, 179, 180-181, 184-185, 186-187(B), 187(T), 189(T), 190-191, 192-193, 195, 198-199, 201

Geoffrey Rogers © Interpet Publishing: *Title page, Contents pages,* 7, 10, 12, 16, 26, 29, 68, 69, 70, 74, 85(B), 87, 89, 91(R), 95(R), 97(R), 103(TR), 105(R), 117, 119(TR), 121(T), 135, 145, 151, 159(TR), 161, 171(TL), 173, 175, 183, 188-189(B), 191, 194, 196-197

Tropical Marine Centre: 32(BL), 33(BR), 35(L)

Computer graphics by Phil Holmes and Stuart Watkinson © Interpet Publishing.

Index compiled by Amanda O'Neill.

The publishers would like to thank the following for their help: Swallow Aquatics at Colchester, East Harling, Rayleigh and Southfleet; Arcadia, Croydon, Surrey; Aquaworld, Warringtom, Cheshire.

Publisher's note